T0222356

This series aims to report new developments in mathematical research and teaching - quickly, informall and at a high level. The type of material considered for publication includes:

1. Preliminary drafts of original papers and monographs
2. Lectures on a new field, or presenting a new angle on a classical field
3. Seminar work-outs
4. Reports of meetings, provided they are
   a) of exceptional interest or
   b) devoted to a single topic.

Texts which are out of print but still in demand may also be considered if they fall within these categorie

The timeliness of a manuscript is more important than its form, which may be unfinished or tentativ Thus, in some instances, proofs may be merely outlined and results presented which have been or wi later be published elsewhere.

Manuscripts should comprise not less than 100 pages.

Publication of *Lecture Notes* is intended as a service to the international mathematical community, i that a commercial publisher, Springer-Verlag, can offer a wider distribution to documents which woul otherwise have a restricted readership. Once published and copyrighted, they can be documented in th scientific literature. -

Manuscripts

Manuscripts are reproduced by a photographic process; they must therefore be typed with extreme care Symbols not on the typewriter should be inserted by hand in indelible black ink. Corrections to the type script should be made by sticking the amended text over the old one, or by obliterating errors with whit correcting fluid. Authors receive 75 free copies.

The typescript is reduced slightly in size during reproduction; best results will not be obtained unless th text on any one page is kept within the overall limit of 18 x 26.5 cm (7 x 10½ inches). The publishers wil be pleased to supply on request special stationery with the typing area outlined.

Manuscripts in English, German or French should be sent to Prof. Dr. A. Dold, Mathematisches Institut de Universität Heidelberg, 69 Heidelberg/Germany, Tiergartenstraße or Prof. Dr. B. Eckmann, Eidgenössi sche Technische Hochschule, CH-8006 Zürich/Switzerland.

Lecture Notes in Physics

Bisher erschienen/Already published

Vol. 1: J. C. Erdmann, Wärmeleitung in Kristallen, theoretische Grundlagen und fortgeschrittene experimentelle Methoden. II, 283 Seiten. 1969. DM 20,–

Vol. 2: K. Hepp, Théorie de la renormalisation. III, 215 pages. 1969. DM 18,–

Vol. 3: A. Martin, Scattering Theory: Unitarity, Analytic and Crossing. IV, 125 pages. 1969. DM 14,–

Vol. 4: G. Ludwig, Deutung des Begriffs physikalische Theorie und axiomatische Grundlegung der Hilbertraumstruktur der Quantenmechanik durch Hauptsätze des Messens. XI, 469 Seiten. 1970. DM 28,–

Vol. 5: M. Schaaf, The Reduction of the Product of Two Irreducible Unitary Representations of the Proper Orthochronous Quantummechanical Poincaré Group. IV, 120 pages. 1970. DM 14,–

Vol. 6: Group Representations in Mathematics and Physics. Edited by V. Bargmann. V, 340 pages. 1970. DM 24,–

Vol. 7: R. Balescu, J. L. Lebowitz, I. Prigogine, P. Résibois, Z. W. Salsburg, Lectures in Statistical Physics. V, 181 pages. 1971. DM 18,–

Vol. 9: D. W. Robinson, The Thermodynamic Pressure in Quantu Statistical Mechanics. V, 115 pages. 1971. DM 14,–

Vol. 10: J. M. Stewart, Non-Equilibrium Relativistic Kinetic Theo III, 113 pages. 1971. DM 14,–

Vol. 11: O. Steinmann, Perturbation Expansions in Axiomatic Fie Theory. III, 126 pages. 1971. DM 14,–

Lecture Notes in Mathematics

A collection of informal reports and seminars
Edited by A. Dold, Heidelberg and B. Eckmann, Zürich

278

Hervé Jacquet

The City University of New York, New York, NY/USA

Automorphic Forms on GL(2)

Part II

AMS Subject Classifications (1970): 10D15

ISBN 3-540-05931-8 Springer-Verlag Berlin · Heidelberg · New York
ISBN 0-387-05931-8 Springer-Verlag New York · Heidelberg · Berlin

This work is subject to copyright. All rights are reserved, whether the whole or part of the material is concerned, specifically those of translation, reprinting, re-use of illustrations, broadcasting, reproduction by photocopying machine or similar means, and storage in data banks.

Under § 54 of the German Copyright Law where copies are made for other than private use, a fee is payable to the publisher, the amount of the fee to be determined by agreement with the publisher.

© by Springer-Verlag Berlin · Heidelberg 1972. Library of Congress Catalog Card Number 76-108338.

Offsetdruck: Julius Beltz, Hemsbach/Bergstr.

if not for a suggestion of his. In particular, the application to quadratic extensions of §20 was, after the oral indications he gave to me, a routine exercise.

I gratefully acknowledge the support of The City University of New York and the National Science Foundation (GP 27952). I wish also to express my thanks to Mrs. Sophie Gerber for typing these notes with competence, patience and understanding.

Finally, I wish to apologize to the mathematical community for presenting a set of notes, if not as bulky, at least as tedious as the first one. My excuse is that I am trying to prove the conjectures outlined by R. Langlands in as many cases as possible. No doubt, the present work will merge into the general case and disappear from the realm of mathematics. I can only hope that somehow, it will sometimes be of some use, however feeble, to the mathematical community.

Hervé Jacquet
January 1972
New York

Introduction

This is a continuation of "Automorphic Forms on GL(2)". Unfortunately, the reader (if any) will have to have a serious knowledge of the two first chapters of the first volume if he is to find his way through the second one. Perhaps reading Godement's "Notes on Jacquet-Langlands", Institute for Advanced Study (1970) will help him in satisfying this stringent requirement. The main purpose of the second volume is to reformulate and extend a classical result: if

$$\sum a_n/n^s \quad , \quad \sum b_n/n^s$$

are two Dirichlet series associated with automorphic forms (in the classical sense) then the Dirichlet series

$$\sum a_n b_n/n^s$$

is convergent in some right half space, can be analytically continued in the whole complex plane as a meromorphic function of s and satisfies a suitable functional equation. Anything novel in this work comes from the point of view which is the theory of group representations. The local theory in §14 to 18 is a preparation of a technical nature for the global theory of § 19. The motivations appear therefore only in the latter section. The reader should read first §14, take for granted the results of §16 to §18 and then go to §19. §20 is an application to quadratic extensions. Again there is nothing really new in it.

In the Bibliography I have tried to indicate my indebtness to previous authors. But I could not however acknowledge completely my indebtness to G. Shimura. This paper would have never been written

Table of Contents

Chapter IV: Local Theory for $GL(2) \times GL(2)$

§14. Existence of a Functional Equation (non-archimedean case) 1

§15. Explicit Computations 23

§16. Explicit Computations (continued) 62

§17. Real Case 70

§18. Complex Case 99

Chapter V: Global Theory for $GL(2) \times GL(2)$

§19. Global Functional Equation for $GL(2) \times GL(2)$ 117

§20. Application to Quadratic Extensions 132

Bibliography 141

Note: This paper is a continuation of "Automorphic Forms on GL(2)", Volume I (Lecture Notes in Mathematics, Volume 114). Unfortunately, the section numbers overlap. Sorry.

Summary and Notations

In general, the notations are the same as in the first volume which is referred to as [1] (see Bibliography). Since the typography is actually different, we give again the principal notations used in [1] as well as a partial list of the new notations introduced in the second volume. We also recall some results which were proved for instance in [8].

The ground field F is a local (commutative) field in Chapter IV (§14 to §18) and an $\underline{A}$-field in Chapter V (§19 and §20). The group G is the group $GL(2)$ regarded as an algebraic group defined over F. We consider the following algebraic subgroups:

$$P = \left\{\begin{pmatrix} a & b \\ 0 & c \end{pmatrix}\right\}, \quad N = \left\{\begin{pmatrix} 1 & b \\ 0 & 1 \end{pmatrix}\right\}, \quad Z = \left\{\begin{pmatrix} a & 0 \\ 0 & a \end{pmatrix}\right\}, \quad A = \left\{\begin{pmatrix} a & 0 \\ 0 & b \end{pmatrix}\right\} .$$

Thus Z_F can be identified to $F^\times$. We also set

$$w = \begin{pmatrix} 0 & 1 \\ -1 & 0 \end{pmatrix}, \quad \eta = \begin{pmatrix} -1 & 0 \\ 0 & 1 \end{pmatrix} .$$

When F is local we denote by ψ_F or ψ a nontrivial additive character of F and let dx be the self dual Haar measure on F. We denote also by α_F or α the module on F, the module of x being denoted $|x|_F$ or simply $|x|$.

In §14 to §16, the ground field is nonarchimedean. We then denote by q the cardinality of the residual field of F and by v_F or v the normalized valuation. Thus $|x|_F = q^{-v(x)}$. We let R be the ring of integers in F and K be the group $GL(2,K)$.

What we call an Euler factor is a function of s (in $\underline{C}$) of the form $P(q^{-s})^{-1}$ where P is a polynomial such that $P(0) = 1$. In [1] or [8] we associate to each quasi-character χ of $F^\times$ a factor

$\epsilon(s,\chi,\psi_F)$ as well as an Euler factor $L(s,\chi)$. In addition we introduce here the notation

$$\epsilon'(s,\chi,\psi_F) = \epsilon(s,\chi,\psi_F)L(1-s,\chi^{-1})/L(s,\chi) .$$

In [1] we have defined the "irreducible admissible representations" of G_F. If π is such a representation and if it is infinite dimensional it has a Kirillov model noted $\mathcal{K}(\pi,\psi_F)$ here (cf. 2.13 in [1]) as well as a Whittaker model $\mathcal{W}(\pi,\psi_F)$ (cf.2.14 in [1]). We also introduce an Euler factor $L(s,\pi)$ as well as a factor $\epsilon(s,\pi,\psi_F)$. In addition, we set

$$\epsilon'(s,\pi,\psi_F) = \epsilon(s,\pi,\psi_F)L(1-s,\tilde{\pi})/L(s,\pi) ,$$

where $\tilde{\pi}$ is the representation contragredient to π.

If χ is a quasi-character of F the representation $\pi \otimes \chi$ is defined as being $\pi(g)\chi(\det g)$.

For all integers n we denote by $\mathcal{S}(F^n)$ the space of Schwarz-Bruhat functions on F^n. We also denote by $\mathcal{S}(F^\times)$ the space of locally constant compactly supported functions on $F^\times$. If Φ belongs to $S(F^2)$ we set

$$g.\Phi(x,y) = \Phi[(x,y)g] , \quad z(\chi,\Phi) = \int \Phi(0,t)\chi(t)d^\times t ,$$

where $d^\times t$ is a multiplicative Haar measure and χ a quasi-character. Let π_i, $i = 1,2$ be an irreducible admissible representation of G_F, ω_i the quasi-character of $Z_F = F^\times$ defined by

$$\pi_i(a) = \omega_i(a)1 ,$$

and $\omega = \omega_1\omega_2$. Assuming π_1 and π_2 to be both infinite dimensional, for $\Phi \in \mathcal{S}(F^2)$, W_i in $W(\pi_i,\psi)$, we set

$$\Psi(s,W_1,W_2,\Phi) = \int_{Z_F N_F \backslash G_F} W_1(g)W_2(\eta g)z(\alpha^{2s}\omega, g.\Phi)\,|\det g|^s\, dg ,$$

$$\tilde{\Psi}(s,W_1,W_2,\Phi) = \int_{Z_F N_F \backslash G_F} W_1(g)W_2(\eta g)z(\alpha^{2s}\omega^{-1}, g.\Phi)\,|\det g|^s \omega^{-1}(\det g)dg.$$

The analytical continuation of those integrals (which are defined for Res large enough) leads to the definition of an Euler factor $L(s,\pi)$ as well as a factor $\epsilon(s,\pi,\psi)$, where π is the external tensor product $\pi_1 \times \pi_2$. We set

$$\epsilon'(s,\pi,\psi) = \epsilon(s,\pi,\psi)L(1-s,\tilde{\pi})/L(s,\pi) .$$

Then we have the functional equation

$$\tilde{\Psi}(1-s,W_1,W_2,\hat{\Phi}) = \omega_2(-1)\epsilon'(s,\pi,\psi)\Psi(s,W_1,W_2,\Phi) ,$$

where $\hat{\Phi}$ is defined by

$$\hat{\Phi}(x,y) = \int \Phi(u,v)\psi(yu-xv)du\, dv \quad .$$

We conjecture that the factors obey the following rule: (notations are as in [1] §12) if $\pi_i = \pi(\sigma_i)$ where σ_i is a two dimensional representation of the Weil group W_F then

$$L(s,\pi) = L(s,\sigma_1 \otimes \sigma_2) , \quad \epsilon(s,\pi,\psi) = \epsilon(s,\sigma_1 \otimes \sigma_2,\psi) ,$$

where the factors in the right-hand side are the ones defined in [9]. Although we fell short of such a goal, all our explicit results are compatible with this assertion (cf. in particular, 19.16).

In §17 the ground field F is $\underline{R}$ the field of real numbers. The group K is the group orthogonal $O(2,\underline{R})$. We have similar notions and results. Of course, we do not consider representations of G_F but rather representations of $\mathcal{H}(G,K)$, the Hecke algebra (cf. [1], §5).

Also we use the following notion of Euler factor. First set

$$G_1(s) = \pi^{-\frac{1}{2}s}\,\Gamma(\tfrac{1}{2}s)\ ,\ G_2(s) = (2\pi)^{1-s}\,\Gamma(s)\ ,$$

where Γ is the gamma function and $\pi = 3.1416.....$. A Euler factor is a function of the form

$$P(s)\prod_i G_1(s+s_i)\prod_j G_2(s+s_j)$$

where P is a polynomial and the s_i , s_j , some constants.

In §18 the ground field F is $\underline{C}$ the field of complex numbers. The group K is the group $U(2,\underline{C})$. We consider again representations of the Hecke algebra $\mathcal{H}(G,K)$. A Euler factor is now a function of the form

$$P(s)\prod_j G_2(s+s_j)$$

where the s_j are some constants and P is a polynomial.

In Chapter V the ground field F is an $\underline{A}$-field. We then follow standard notations and denote by $\underline{A}$ the ring of adèles and $\underline{I}$ the group of idèles; if v is a place of F then F_v is the corresponding local field and $G_v = GL(2,F_v)$; K_v is the standard maximal compact subgroup of G_v and $K = \prod K_v$. Let ω_i , $i = 1,2$, be two quasi-characters of $\underline{I}/F^\times$ and π_i an admissible irreducible representation of the Hecke algebra $\mathcal{H}(G_{\underline{A}},K)$ which is contained in the space $\mathcal{G}_0(\omega_i)$ (space of cusp forms). Then there is a global Whittaker model $\mathcal{W}(\pi_i,\psi)$ (cf. [1], 9.2). For Φ in $\mathcal{S}(\underline{A}^2)$, W_i in $\mathcal{W}(\pi_i,\psi)$ we define as in the local case, two integrals $\Psi(s,W_1,W_2,\Phi)$ and $\tilde{\Psi}(s,W_1,W_2,\Phi)$. They are convergent for $\mathrm{Re}\,s$ large enough, can be analytically continued as a meromorphic function of s and satisfy the functional equation

$$\Psi(s,W_1,W_2,\Phi) = \tilde{\Psi}(1-s,W_1,W_2,\hat{\Phi}) ,$$

where $\hat{\Phi}$ is defined by

$$\hat{\Phi}(x,y) = \int \Phi(u,v)\psi(uy-xv)du\,dv \quad .$$

Here ψ is a nontrivial additive character of $\underline{A}/F$ and du or dv the self dual Haar measure on $\underline{A}$. (Cf. 19.13).

Now we summarize some notations and results used in §19 without further references. Let F_0 and F_1 be two continuous positive functions on $\underline{R}_+^\times$ satisfying the following conditions:

$$F_0 + F_1 = 1 \ , \ F_1(t) = \check{F}_0(t) \quad (= F_0(t^{-1})) \ ,$$

there are t_0 and t_1 such that $0 < t_0 < 1 < t_1$ and

$$F_0(t) = 0 \quad \text{for} \quad 0 < t < t_0 \quad \text{and} \quad F_0(t) = 1 \quad \text{for} \quad t_1 < t \ .$$

If ω is a quasi-character of $\underline{I}/F^\times$ we set

$$\lambda(\omega) = \int_{\underline{I}/F^\times} \omega(a)F_1(|a|)\,da$$

where da is a Haar measure on $\underline{I}/F^\times$. Then if $|\omega| = \alpha_F^s$ with $s > 0$ (α_F denoting the module on $\underline{I}$) the integral is convergent. It is a meromorphic function of ω and $\lambda(\omega) + \lambda(\omega^{-1}) = 0$. The only pole is simple and occurs for $\omega = 1$. If Φ belongs to $\mathcal{S}(\underline{A}^2)$, we define

$$\theta^0(\omega,\Phi) = \int_{\underline{I}/F^\times} \sum_{(\xi,\eta)\neq(0,0)} \Phi(a(\xi,\eta))\omega(a)F_0(|a|)\,da \ ,$$

$$\theta^1(\omega,\Phi) = \int_{\underline{I}/F^\times} \sum_{(\xi,\eta)\neq(0,0)} \Phi(a(\xi,\eta))\omega(a)F_1(|a|)\,da \ ,$$

the sum being extended to all pairs in F^2 except the pair $(0,0)$.

Then $\theta^0(\omega,\Phi)$ is convergent for all ω and $\theta^1(\omega,\Phi)$ for $|\omega| = \alpha_F^s$ with $s > 2$. Moreover, by Poisson formula,

$$\theta^1(\omega,\Phi) = \theta(\alpha_F^2\omega^{-1},\hat{\Phi}) - \lambda(\alpha_F^2\omega^{-1})\hat{\Phi}(0) - \lambda(\omega)\Phi(0) .$$

Finally the function

$$g \longrightarrow \theta^0(\omega,g.\Phi)$$

where $g.\Phi(x,y) = \Phi((x,y)g)$ is, with the terminology of [1], a slowly increasing function on $G_F\backslash G_{\underline{A}}$. (Cf. [7], VII §5 and [8] §11). We also introduce the subgroup G_0 of g in $G_{\underline{A}}$ such that $|\det g| = 1$ and the subset G'. If F is a number field $G' = G_0$. If F is a function field and the field of constant has cardinality Q we select g_1 such that $|\det g_1| = Q^{-1}$ and set $G' = G_0 \cup G_0 g_1$. Then in both cases $G_{\underline{A}} = Z_{\underline{A}}G'$.

Combining the local and global results for the integrals Ψ and $\tilde{\Psi}$ in the customary fashion, we arrive at the following result which is the main result of the second volume: the representation $\pi = \pi_1 \times \pi_2$ being as above, we set

$$L(s,\pi) = \prod L(s,\pi_v) , \quad \epsilon(s,\pi) = \prod \epsilon(s,\pi_v,\psi_v)$$

where π_v (resp. ψ_v) is the local component of π (resp. ψ, a basic character of $\underline{A}/F$) at the place v; then $L(s,\pi)$ is absolutely convergent in some right half space, can be analytically continued as a meromorphic function of s in the whole complex plane and satisfy the functional equation

$$L(s,\pi) = \epsilon(s,\pi)L(1-s,\tilde{\pi})$$

where $\tilde{\pi}$ is the representation contragredient to π.

XIII

In §20 we let K be a separable quadratic extension of F and denote by $F_{\underline{A}}$ (resp. $K_{\underline{A}}$) the ring of adèles of F (resp. K) and $F_{\underline{A}}^{\times}$ (resp. $K_{\underline{A}}^{\times}$) the group of idèles. If π is now an "irreducible representation" of $GL(2,F_{\underline{A}})$ contained in the space of cusp forms, we associate with π an irreducible representation σ of $GL(2,K_{\underline{A}})$ and show, more or less, that σ is contained in the space of automorphic forms for $GL(2,K_{\underline{A}})$. If w is a place of K above the place v of F, the representation σ_w should be obtained in terms of the representation π_v according to the following rule: assume that $\pi_v = \pi(\tau)$ where τ is a two dimensional representation of the Weil group W_{F_v}, then $\sigma_w = \pi(\tau')$ where τ' is the restriction of τ to the subgroup W_{K_w} of W_{F_v}. Such is the case if v is archimedean. If v is nonarchimedean, we do not know at the moment that all representation of G_v are associated with representation of W_{F_v}, (actually the special representation is not). So we have to introduce an "ad hoc" notion.

Chapter IV: Local Theory for GL(2) x GL(2)

§14. Existence of a functional equation (non-archimedean case)

In Chapters IV and V the group G is the group $GL(2)$ regarded as an algebraic group over the ground field F. In §§14, 15, 16 the field F is local and non-archimedean.

We assume the reader to be familiar with the notations and results of [1] Chapter I. (See also Summary and Notations).

Let π be an irreducible representation of G_F on a complex vector space V. We assume that π is infinite dimensional. Once a non-trivial additive character ψ of F has been chosen, the representation π has a unique Kirillov model $\mathcal{K}(\pi,\psi)$. It is characterized by the following conditions. It is a space of complex valued functions on $F^{\times}$ and for a in $F^{\times}$, x in $F^{\times}$ and φ in $\mathcal{K}(\pi,\psi)$, the relations

$$(14.1) \qquad \pi\begin{pmatrix} a & 0 \\ 0 & 1 \end{pmatrix}\varphi(b) = \varphi(ab) \quad \text{for all } a \text{ in } F^{\times}$$

$$\pi\begin{pmatrix} 1 & x \\ 0 & 1 \end{pmatrix}\varphi(b) = \varphi(b)\psi(bx)$$

hold. In general, we denote by B_F the group of all matrices of the form

$$\begin{pmatrix} a & x \\ 0 & 1 \end{pmatrix}, \ a \in F^{\times}, \ x \in F$$

and by ξ the representation of B_F on the space of all complex valued functions on $F^{\times}$ which is defined by (14.1).

To determine the space $\mathcal{K}(\pi,\psi)$ we shall use the following simple lemma:

Lemma 14.2: *Let f be a function in $\mathcal{S}(F)$ and μ a quasi-character* of $F^\times$. Let also i be one of the integers $0,1,2$. Then the integral

$$\int f(a)\, v^i(a)\, |a|^s\, \mu(a)\, d^\times a$$

is absolutely convergent if Res is large enough. As a function of s , it is, in fact, a rational function of q^{-s} .

If the function f vanishes at $x = 0$ the integral is convergent for all s and defines a finite Laurent series in q^{-s} . So we may assume that $f = \Phi_0$ the characteristic function of R the ring of integers in F . Then, if μ is trivial, we find

$$\int \Phi_0(a)\, |a|^s\, v^i(a)\, d^\times a = \sum_{n \geq 0} q^{-ns}\, n^i$$

both sides being simultaneously defined. As the right-hand side is absolutely convergent if Res > 0 so is the left-hand side. Hence the first assertion of the lemma.

Still assuming $f = \Phi_0$ we find that the integral of the lemma vanishes if μ is ramified. So for the last assertion of the lemma, we may assume μ to be unramified, or even trivial and also that $f = \Phi_0$. In that case, the integral assumes the following values:

$$(14.2.1) \qquad \begin{array}{ll} (1 - X)^{-1} & \text{if } i = 0, \\ X(1 - X)^{-2} & \text{if } i = 1, \\ X(1 + X)(1 - X)^{-3} & \text{if } i = 2, \end{array}$$

where $X = q^{-s}$. The lemma follows.

From [2], we borrow the following lemma.

Lemma 14.3: The space $\mathcal{K}(\pi,\psi)$ consists of

(1) all locally constant, compactly supported functions on $F^\times$ if the

<u>representation</u> π <u>is absolutely cuspidal</u>; (<u>space denoted</u> $\mathcal{S}(F^{\times})$) ,

(2) <u>all functions of the form</u>

$$a \longmapsto f(a)\,\mu(a)\,|a|^{\frac{1}{2}} + g(a)\,\nu(a)\,|a|^{\frac{1}{2}}$$

<u>where</u> f <u>and</u> g <u>belong to</u> $\mathcal{S}(F)$ <u>if</u> $\pi = \pi(\mu,\nu)$ <u>and</u> $\mu \neq \nu$;

(3) <u>all functions of the form</u>

$$a \longmapsto f(a)\,\mu(a)\,|a|^{\frac{1}{2}} + g(a)\,\mu(a)v(a)\,|a|^{\frac{1}{2}}$$

<u>where</u> f <u>and</u> g <u>belong to</u> $\mathcal{S}(F)$ <u>if</u> $\pi = \pi(\mu,\mu)$;

(4) <u>all functions of the form</u>

$$a \longmapsto f(a)\,\mu(a)\,|a|^{\frac{1}{2}} \quad ,$$

<u>where</u> f <u>belongs to</u> $\mathcal{S}(F)$, <u>if</u> $\pi = \sigma(\mu,\nu)$ <u>and</u> $\mu.\nu^{-1} = \alpha_F$.

First, we remind ourselves of the definition of the factor $L(s,\mu)$ when μ is a quasi-character of $F^{\times}$. It is such that for all f in $\mathcal{S}(F)$

$$\int f(a)\,|a|^{s}\,\mu(a)\,d^{\times}a = P(s)L(s,\pi)$$

where P is a polynomial in q^{-s} and q^{s} . Conversely, given such a polynomial P , there is an element f of $\mathcal{S}(F)$ such that the above relation is true. Moreover, if we impose to f to satisfy

$$f(\epsilon a) = f(a)\,\mu^{-1}(\epsilon) \quad \text{for all} \quad \epsilon \quad \text{in} \quad R^{\times} ,$$

then f is unique.

Similarly, the factor $L(s,\pi)$ satisfies the following conditions. For all φ in $\mathcal{K}(\pi,\psi)$

$$\int \varphi(a)\,|a|^{s-\frac{1}{2}}\,d^{\times}a = P(s)L(s,\pi)$$

where $P \in \underline{C}[q^{-s},q^{s}]$. Conversely, given P , there is φ in $\mathcal{K}(\pi,\psi)$ such that the above relation is true and it is unique, if we impose to φ to satisfy

(14.3.5) $$\varphi(\epsilon a) = \varphi(a) \quad \text{for all} \quad \epsilon \in R^{\times} .$$

So, knowing $L(s,\pi)$, we can actually compute the space $\mathcal{K}_0(\pi,\psi)$ of all φ in $\mathcal{K}(\pi,\psi)$ which satisfy (14.3.5). Now $L(s,\pi)$ is given by the following table:

case number in (14.3)	$L(s,\pi)$
1	1
2	$L(s,\mu)L(s,\nu)$
3	$L(s,\mu)^2$
4	$L(s,\mu)$.

Therefore, the determination of $\mathcal{K}_0(\pi,\psi)$ is an easy matter. One determines the whole space $\mathcal{K}(\pi,\psi)$ by replacing π by the tensor product $\pi \otimes \chi$. The lemma follows.

Let π now be an admissible and irreducible representation of $G_F \times G_F$. Then $\pi = \pi_1 \times \pi_2$ where, for $i = 1,2$, π_i is an admissible irreducible representation of G_F . We assume that neither π_1 nor π_2 is one dimensional.

__Proposition 14.4__: __For__ φ_i __in__ $\mathcal{K}(\pi_i,\psi)$, $i = 1,2$, __we set__:

$$\beta_s(\varphi_1,\varphi_2) = \int \varphi_1(a)\varphi_2(-a)|a|^{s-1} d^{\times}a \quad .$$

__Then there is__ $s_0 \in \underline{R}$ __so that for__ $\operatorname{Re} s > s_0$ __all these integrals are absolutely convergent__. __They are rational functions of__ q^{-s} __and can be written as quotient of an element of__ $\underline{C}[q^{-s},q^{s}]$ __by a fixed polynomial__ *in* $\underline{C}[q^{-s}]$ *(the same for all* φ_1 *and* φ_2*).*

This follows at once from (14.2) and (14.3).

In what follows, "for almost all s" will mean for all s except those s which belong to a set Y such that q^{-Y} is finite. Then, by analytic continuation, the bilinear form β_s is defined for "almost all s" and satisfies the identity

$$\beta_s\left[\xi\begin{pmatrix} a & x \\ 0 & 1\end{pmatrix}\varphi_1 , \xi\begin{pmatrix} a & x \\ 0 & 1\end{pmatrix}\varphi_2\right] = |a|^{1-s}\, \beta_s(\varphi_1,\varphi_2)$$

for all a in $F^\times$ and x in F and φ_i in $\mathcal{K}(\pi_i,\psi)$.

In general, consider a bilinear form γ on the product $\mathcal{K}(\pi_1,\psi) \times \mathcal{K}(\pi_2,\psi)$ such that

$$\gamma\left[\xi\begin{pmatrix} a & x \\ 0 & 1\end{pmatrix}\varphi_1 , \xi\begin{pmatrix} a & x \\ 0 & 1\end{pmatrix}\varphi_2\right] = |a|^{1-s}\, \gamma(\varphi_1,\varphi_2) \tag{14.4.1}$$

for all a in $F^\times$ and x in F and φ_i in $\mathcal{K}(\pi_i,\psi)$.

Lemma 14.4.2: For almost all s, *every bilinear form* γ *which satisfies the relation* (14.4.1) *is proportional to* β_s.

We shall use a series of lemmas.

Lemma 14.4.3: Let γ *be a bilinear form on* $\mathcal{S}(F^\times) \times \mathcal{S}(F^\times)$ *which satisfies the relation* (14.4.1). *Then* γ *is proportional to* β_s *(whose restriction to* $\mathcal{S}(F^\times) \times \mathcal{S}(F^\times)$ *is always defined).*

As the representation ξ of B_F on $\mathcal{S}(F^\times)$ is irreducible ([1] Lemma (2.9.1)) it is enough to show that there is a constant c so that for every φ in $\mathcal{S}(F^\times)$

$$\gamma(\varphi,\varphi_0) = c\beta_s(\varphi,\varphi_0)$$

where we denote by φ_0 the characteristic function of $R^\times$.

This relation can also be written as

$$\gamma(\varphi,\varphi_0) = c \int_{R^\times} \varphi(\epsilon)d\epsilon$$

where $d\epsilon$ is the normalized Haar measure on $R^\times$. Now for ϵ in $R^\times$ we have

$$\xi\begin{pmatrix}\epsilon & 0\\ 0 & 1\end{pmatrix}\varphi_0 = \varphi_0 .$$

Hence

$$\gamma\left[\xi\begin{pmatrix}\epsilon & 0\\ 0 & 1\end{pmatrix}\varphi,\varphi_0\right] = \gamma(\varphi,\varphi_0) .$$

This implies that there are constants a_n, $n \in \underline{Z}$, so that

$$(*) \qquad \gamma(\varphi,\varphi_0) = \sum_{n\in\underline{Z}} a_n \int_{R^\times} \varphi(\epsilon\varpi^n)d\epsilon \quad \text{for all } \varphi \text{ in } \mathcal{S}(F^\times) .$$

(For a given φ in $\mathcal{S}(F^\times)$ only a finite number of terms in the right-hand side do not vanish). It will be enough to show that all the constants a_n but a_0 vanish.

Certainly, there is no harm in assuming the character ψ to be of order zero. Then if

$$u = \begin{pmatrix}1 & 1\\ 0 & 1\end{pmatrix}$$

we have

$$\xi(u)\varphi_0 = \varphi_0 ,$$

hance also

$$\gamma(\xi(u)\varphi,\varphi_0) = \gamma(\varphi,\varphi_0) .$$

Applying this to the characteristic function of $\varpi^{-m}R^\times$ with $m > 0$

and taking in account the relation (*) we find:

$$a_{-m}\int_{R^\times}\psi(\varepsilon\,\varpi^{-m})d\varepsilon = a_{-m}\ .$$

If $m > 1$ the left-hand side vanishes and we find $a_{-m} = 0$. If $m = -1$, we find

$$a_{-1}(1-q)^{-1} = a_{-1}$$

and again $a_{-1} = 0$.

The relation

$$\int_{\varpi^{-1}R} \xi\begin{pmatrix}1 & x\\ 0 & 1\end{pmatrix}\varphi_0\,dx = 0$$

is immediate. It implies

$$\gamma\left[\int_{\varpi^{-1}R} \xi\begin{pmatrix}1 & x\\ 0 & 1\end{pmatrix}\varphi dx, \varphi_0\right] = 0\ .$$

Applying this to the characteristic function of $\varpi^m R^\times$, with $m > 1$, and taking in account the relation (*) we find,

$$a_m = 0\ ,\quad \text{if}\quad m > 1\ .$$

This completes the proof of Lemma (14.4.3).

Note that, if the representations π_1 and π_2 are both absolutely cuspidal, Lemma (14.4.2) reduces, in fact, to Lemma (14.4.3). Similarly, if π_2, for instance, is absolutely cuspidal, Lemma (14.4.2) will reduce to the following lemma.

<u>Lemma 14.4.4</u>: <u>Let</u> γ <u>be a bilinear form on</u> $\mathcal{K}(\pi_1,\psi)\times\mathcal{S}(F^\times)$ <u>which satisfies the relation</u> (14.4.1). <u>Then</u> γ <u>is proportional to</u> β_s.

By the previous lemma we already know that there is a constant c so that

$$\gamma(\varphi_1,\varphi_2) = c\beta_s(\varphi_1,\varphi_2)$$

for φ_1 and φ_2 in $\mathcal{S}(F^\times)$. Using again the irreducibility of $\mathcal{S}(F^\times)$ under B_F, we see that it will be enough to show that there is a non-zero element φ_2 of $\mathcal{S}(F^\times)$ so that

$$\gamma(\varphi_1,\varphi_2) = c\beta_s(\varphi_1,\varphi_2)$$

for all φ_1 in the Kirillov model of π_1.

Set

$$u = \begin{pmatrix} 1 & x \\ 0 & 1 \end{pmatrix}$$

then

$$\gamma(\varphi_1,\varphi_2 - \xi(u^{-1})\varphi_2) = \gamma(\varphi_1 - \xi(u)\varphi_1,\varphi_2)$$

for φ_1 in $\mathcal{K}(\pi_1,\psi)$ and φ_2 in $\mathcal{S}(F^\times)$. There is a similar formula for β_s. As the difference

$$\varphi_1 - \xi(u)\varphi_1$$

belongs to $\mathcal{S}(F^\times)$ we find that

$$\gamma(\varphi_1,\varphi_2 - \xi(u^{-1})\varphi_2) = c\beta_s(\varphi_1,\varphi_2 - \xi(u^{-1})\varphi_2)$$

for all φ_1 in $\mathcal{K}(\pi_1,\psi)$. As we may choose φ_2 and u in such a way that

$$\varphi_2 - \xi(u^{-1})\varphi_2 \neq 0$$

the lemma is proved.

Now we take the proof of (14.4.2). We exclude first the values of s for which β_s is not defined. From (14.4.4) we already know that there is a constant c so that

(**) $$\gamma(\varphi_1,\varphi_2) = c\beta_s(\varphi_1,\varphi_2)$$

if φ_1 and φ_2 belong to $\mathcal{K}(\pi_1,\psi)$ and $\mathcal{K}(\pi_2,\psi)$ respectively and at least one of those functions belong to $\mathcal{S}(F^\times)$. We have to prove that the same relation is true for all pairs (φ_1,φ_2) in $\mathcal{K}(\pi_1,\psi) \times \mathcal{K}(\pi_2,\psi)$. By linearity, we have only to show this for each one of the following cases:

(14.4.5) $$\varphi_1(a) = \mu_1(a)f_1(a) \ , \ \varphi_2(a) = \mu_2(a)f_2(a) \ ,$$

(14.4.6) $$\varphi_1(a) = \mu_1(a)f_1(a)v(a) \ , \ \varphi_2(a) = \mu_2(a)f_2(a) \ ,$$

(14.4.7) $$\varphi_1(a) = \mu_1(a)f_1(a)v(a) \ , \ \varphi_2(a) = \mu_2(a)f_2(a)v(a) \ .$$

Here μ_1 and μ_2 are quasi-characters and f_1, f_2 belong to $\mathcal{S}(F)$.

In case (14.4.5) we choose a in $F^\times$ and set

$$\xi\begin{pmatrix} a & 0 \\ 0 & 1 \end{pmatrix} \varphi_i = \varphi_i' + \mu_i(a)\varphi_i \quad \text{for } i = 1,2 \ .$$

Then $\varphi_i'(x)$ vanishes if x is small enough. So φ_i' belongs to $\mathcal{S}(F^\times)$ and we know that the relation (**) is true if we substitute φ_1' to φ_1 and/or φ_2' to φ_2. Now the assumptions on γ imply that

$$(|a|^{1-s} - \mu_1\mu_2(a))\ \gamma(\varphi_1,\varphi_2) = \gamma(\varphi_1',\varphi_2') + \mu_2(a)\gamma(\varphi_1',\varphi_2) + \mu_1(a)\gamma(\varphi_1,\varphi'_2) \ .$$

There is a similar relation for β_s. We exclude the values of s for which $a^{1-s} = \mu_1\mu_2$. Then we can choose a so that

$$|a|^{1-s} \neq \mu_1\mu_2(a) \ .$$

Comparing the relations for γ and β_s we find

$$\gamma(\varphi_1,\varphi_2) = c\beta_s(\varphi_1,\varphi_2) \ .$$

Let us take the case (14.4.6). Then the function

$$\psi_1(a) = \mu_1(a)f_1(a)$$

belongs to $\mathcal{K}(\pi_1,\psi)$. For a in $F^\times$ we set

$$\xi\begin{pmatrix} a & 0 \\ 0 & 1 \end{pmatrix}\varphi_1 = \varphi_1' + \mu_1(a)\varphi_1 + \mu_1(a)v(a)\psi_1$$

$$\xi\begin{pmatrix} a & 0 \\ 0 & 1 \end{pmatrix}\varphi_2 = \varphi_2' + \mu_2(a)\varphi_2 \quad .$$

Then φ_1' and φ_2' both belong to $\mathcal{S}(F^\times)$. The assumptions on γ give

$$(|a|^{1-s} - \mu_1\mu_2(a))\gamma(\varphi_1,\varphi_2) = \gamma(\varphi_1',\varphi_2') + \mu_1(a)\gamma(\varphi_1,\varphi_2') + \mu_2(a)\gamma(\varphi_1',\varphi_2)$$
$$+ \mu_1(a)v(a)\gamma(\psi_1,\varphi_2') + \mu_1\mu_2(a)v(a)\gamma(\psi_1,\varphi_2) \ .$$

We again exclude the s for which $a^{1-s} = \mu_1\mu_2$. Then the relation (**) holds for all pairs on the right-hand side (by the previous case). Since the same relation is true for β_s instead of γ we find that (**) holds for the pair (φ_1,φ_2).

In case (14.4.7) we set

$$\psi_i(a) = \mu_i(a)f_i(a)$$

and

$$\xi\begin{pmatrix} a & 0 \\ 0 & 1 \end{pmatrix}\varphi_i = \varphi_i' + \mu_i(a)\varphi_i + \mu_i(a)v(a)\psi_i \quad .$$

Then ψ_i is in $\mathcal{K}(\pi_i,\psi)$ and φ_i' in $\mathcal{S}(F^\times)$. The assumptions on the bilinear form γ give

$$(|a|^{1-s} - \mu_1\mu_2(a))\gamma(\varphi_1,\varphi_2) = \gamma(\varphi_1',\varphi_2') + \mu_1(a)\gamma(\varphi_1,\varphi_2')$$
$$+ \mu_2(a)\gamma(\varphi_1',\varphi_2) + \mu_1(a)v(a)\gamma(\psi_1,\varphi_2')$$

$$+\mu_2(a)\nu(a)\gamma(\varphi_1',\psi_2)+\mu_1\mu_2(a)\nu(a)\gamma(\psi_1,\varphi_2)$$
$$+\mu_1\mu_2(a)\nu(a)\gamma(\varphi_1,\psi_2)+\mu_1\mu_2(a)\nu(a)^2\,\gamma(\psi_1,\psi_2)\quad .$$

Again we exclude those s for which $\alpha^{1-s}=\mu_1\mu_2$. Then the relation (**) holds for all pairs on the right-hand side of the above formula. Since the formula is also true for γ_s we find at last that (**) is true for the pair (φ_1,φ_2). This concludes the proof of (14.4.2).

We pass now to the Whittaker model of the representations π_i. The space $\mathfrak{W}(\pi_i,\psi)$ is a space of functions W on G_F which transform on the left according to

$$W\left[\begin{pmatrix}1 & x\\ 0 & 1\end{pmatrix}g\right]=\psi(x)W(g)\ .$$

The space $W(\pi_i,\psi)$ is invariant under right translations and the representation of G_F on it is equivalent to the representation π_i. We set

$$\eta=\begin{pmatrix}-1 & 0\\ 0 & 1\end{pmatrix}\ .$$

Then for W_i in $W(\pi_i,\psi)$ the function

$$g\longmapsto W_1(g)W_2(\eta g)$$

is invariant on the left under the group

$$N_F=\left\{\begin{pmatrix}1 & x\\ 0 & 1\end{pmatrix}\ \middle|\ x\in F\right\}\ .$$

Moreover, if ω_i and ω are the quasi-characters of $F^\times$ defined by

$$\pi_i(a)=\omega_i(a)\quad\text{for } a\in F^\times\ ,\ \omega=\omega_1\omega_2$$

we see that

$$W_1(ga)W_2(\eta ga)=W_1(g)W_2(\eta g)\omega(a)\quad\text{for all } a\in F^\times=Z_F\ .$$

With the notations of [1] on $\mathcal{S}(F^2)$ the distribution

$$z(\alpha^{2s}\omega,\Phi) = \int \Phi(0,t)\,|t|^{2s}\omega(t)d^{\times}t$$

is defined by a convergent integral for Res large enough, and by analytic continuation for all s such that $\alpha^{2s}\omega \neq 1$. So for a given Φ the functions

$$f(g) = z(\alpha^{2s}\omega, g.\Phi)\,|\det g|^s \ , \ h(g) = z(\alpha^{2s}\omega^{-1}, g.\Phi)\,|\det g|^s\,\omega^{-1}(\det g)$$

can be defined for almost all s. When defined they satisfy

$$f\left[\begin{pmatrix} a & x \\ 0 & b \end{pmatrix} g\right] = |ab^{-1}|^s\,\omega^{-1}(b)f(g)\ ,$$

$$h\left[\begin{pmatrix} a & x \\ 0 & b \end{pmatrix} g\right] = |ab^{-1}|^s\,\omega^{-1}(a)h(g)\ .$$

In other words, with the notations of [1], the function f belongs to $\mathcal{B}(\alpha^{s-\frac{1}{2}}, \alpha^{\frac{1}{2}-s}\,\omega^{-1})$ and the function h to $\mathcal{B}(\alpha^{s-\frac{1}{2}}\,\omega^{-1}, \alpha^{\frac{1}{2}-s})$. In particular, the functions

$$g \longmapsto W_1(g)W_2(\eta g)f(g) \quad\text{and}\quad g \longmapsto W_1(g)W_2(\eta g)h(g)$$

are invariant on the left under the group $Z_F N_F$. As G_F and $Z_F N_F$ are unimodular groups, there is an invariant measure dg on the quotient $Z_F N_F \backslash G_F$. We set

$$\Psi(s,W_1,W_2,\Phi) = \int_{ZN\backslash G} W_1(g)W_2(\eta g)f(g)dg\ , \tag{14.5}$$

$$\tilde{\Psi}(s,W_1,W_2,\Phi) = \int_{ZN\backslash G} W_1(g)W_2(\eta g)h(g)dg\ , \tag{14.6}$$

for all W_i in $\mathcal{W}(\pi_i,\psi)$ and Φ in $\mathcal{S}(F^2)$.

Let us substitute the representation $\tilde{\pi}_i = \pi_i \otimes \omega_i^{-1}$ to the representation π_i . Then ω is replaced by ω^{-1} . Now if we substitute to the function $W_i \in \mathcal{W}(\pi_i,\psi)$ the function

$$g \longmapsto W_i(g)\omega_i^{-1}(\det g)$$

which is in $\mathcal{W}(\tilde{\pi}_i,\psi)$, we see that the roles of the integrals (14.5) and (14.6) are exchanged.

The previous results can be translated in terms of these new integrals.

Theorem 14.7: (1) There is s_0 so that for $\text{Re}\, s > s_0$ the functions f and h are defined and the integrals (14.5) and (14.6) absolutely convergent.

(2) (14.5) and (14.6) are rational function of q^{-s} . More precisely, they can be written as quotient of an element of $\underline{C}[q^{-s},q^{s}]$ by a fixed element of $\underline{C}[q^{-s}]$ (which is independent from W_i in $W(\pi_i,\psi)$ and Φ in $\mathcal{S}(F^2)$).

(3) There is a rational function of q^{-s} noted $\gamma(s)$ such that, for all W_i in $\mathcal{W}(\pi_i,\psi)$ and Φ in $\mathcal{S}(F^2)$

$$\tilde{\Psi}(1-s,W_1,W_2,\hat{\Phi}) = \gamma(s)\Psi(s,W_1,W_2,\Phi)$$

where, for Φ in $\mathcal{S}(F^2)$, we denote by $\hat{\Phi}$ the element of $\mathcal{S}(F^2)$ defined by

$$\hat{\Phi}(x,y) = \iint \Phi(u,v)\psi(uy-vx)\,du\,dv \quad .$$

Let K be the group $GL(2,R)$. Since $G_F = Z_F N_F A_F K$ we may compute the integrals in the following manner:

$$\Psi(s,W_1,W_2,\Phi) = \int_{F^\times \times K} W_1\left[\begin{pmatrix} a & 0 \\ 0 & 1\end{pmatrix}k\right] W_2\left[\begin{pmatrix} -a & 0 \\ 0 & 1\end{pmatrix}k\right] |a|^{s-1} z(\alpha^{2s}\omega, k.\Phi)\, d^\times a\, dk \quad ,$$

$$\tilde{\Psi}(s,W_1,W_2,\Phi) = \int_{F^\times \times K} W_1\left[\begin{pmatrix} a & 0 \\ 0 & 1\end{pmatrix}k\right] W_2\left[\begin{pmatrix} -a & 0 \\ 0 & 1\end{pmatrix}k\right] |a|^{s-1}\omega^{-1}(a) z(\alpha^{2s}\omega^{-1}, k.\Phi)$$
$$\times\ \omega^{-1}(\det k)\, d^\times a\, dk \quad .$$

Since the functions to be integrated are K-finite on the right and

$$a \longmapsto W_i\left[\begin{pmatrix} a & 0 \\ 0 & 1\end{pmatrix}k\right]$$

belongs to $\mathcal{K}(\pi_i,\psi)$ for all k , we see that the assertions (1) and (2) are little more than a reformulation of (14.4).

It is more difficult to see that the functional equation of assertion (3) is, in fact, a consequence of (14.4.2). Our starting point is the following lemma.

<u>Lemma 14.7.1</u>: <u>There is</u> s_0 <u>in</u> $\underline{R}$ <u>so that for all</u> s <u>with</u> $\mathrm{Re}\, s > s_0$ <u>and all</u> Φ <u>in</u> $\mathcal{S}(F^2)$, <u>the relation</u>

$$z(\alpha^{2s}\omega, g.\Phi) = 0 \quad \underline{\text{for all}}\ \ g\ \ \underline{\text{in}}\ \ G_F$$

<u>inplies the relation</u>

$$z(\alpha^{2-2s}\omega^{-1}, g.\hat{\Phi}) = 0 \quad \underline{\text{for all}}\ \ g\ \ \underline{\text{in}}\ \ G_F \ .$$

Set

$$w = \begin{pmatrix} 0 & 1 \\ -1 & 0\end{pmatrix} , \quad f(g) = z(\alpha^{2s}\omega, g.\Phi)\, |\det g|^s ,$$
$$h(g) = z(\alpha^{2-2s}\omega^{-1}, g.\hat{\Phi})\, |\det g|^{1-s}\omega^{-1}(\det g) \ .$$

It is enough to see that for $\mathrm{Re}\, s$ large enough the integral

$$\int f\left[w\begin{pmatrix} 1 & x \\ 0 & 1\end{pmatrix}g\right]dx$$

is convergent and equal to

$$\omega(-1)\epsilon'(2s-1,\omega,\psi)^{-1}\ h(g) \quad .$$

Replacing Φ by $g.\Phi$ we see that it is enough to prove this for $g = e$. Then

$$\int f\left[w\begin{pmatrix}1 & x\\ 0 & 1\end{pmatrix}\right]dx = \int dx \int \Phi\left[(0,t)w\begin{pmatrix}1 & x\\ 0 & 1\end{pmatrix}\right]|t|^{2s}\omega(t)d^{\times}t$$

$$= \iint \Phi(-t,-tx)\,|t|^{2s}\omega(t)d^{\times}tdx$$

$$= \omega(-1)\iint \Phi(t,x)\,|t|^{2s-1}\omega(t)d^{\times}tdx \quad .$$

Since the last double integral is absolutely convergent for Res larger than some s_0, we see that for $\mathrm{Res} > s_0$ the first integral is convergent and equal to the last one. The last integral in turn can be written as

$$\omega(-1)Z(\varphi,2s-1,\omega) \ ,$$

where φ is the Schwartz-Bruhat function in one variable defined by

$$\varphi(t) = \int \Phi(t,x)dx \ .$$

Using the local functional equation we see that this is

$$\omega(-1)\epsilon'(2s-1,\omega,\psi)^{-1}\ Z(\hat{\varphi},2-2s,\omega^{-1})$$

where

$$\hat{\varphi}(t) = \int \psi(yt)dy\varphi(y) = \int \psi(yt)dy\int\Phi(y,x)dx = \hat{\Phi}(0,t) \ .$$

Now to prove the required identity and the lemma we have only to observe that

$$Z(\hat{\varphi},2-2s,\omega^{-1}) = z(\alpha^{2-2s}\omega^{-1},\hat{\Phi}) = h(e) \ .$$

In addition to the bilinear form β_s , on the product $\mathcal{K}(\pi_1,\psi)\times$

$\mathcal{K}(\pi_2,\psi)$, we introduce the bilinear form

$$\tilde{\beta}_s(\varphi_1,\varphi_2) = \int \varphi_1(a)\varphi_2(-a)|a|^{s-1}\omega^{-1}(a)d^\times a .$$

Its properties are similar to the properties of β_s . In particular, it is defined by a convergent integral for Res large enough and, by analytic continuation, for almost all s . Keeping the notations of the proof of Lemma (14.7.1) we see that for almost all s

$$(14.7.2) \quad \Psi(s,W_1,W_2,\Phi) = \int_K f(k)dk\ \beta_s(\pi_1(k)\varphi_1,\pi_2(k)\varphi_2) ,$$

$$(14.7.3) \quad \tilde{\Psi}(1-s,W_1,W_2,\hat{\Phi}) = \int_K h(k)dk\ \tilde{\beta}_{1-s}(\pi_1(k)\varphi_1,\pi_2(k)\varphi_2) ,$$

where φ_i is the function

$$a \longmapsto W_i\begin{pmatrix} a & 0 \\ 0 & 1\end{pmatrix} .$$

Now it follows from Proposition (3.2) of [1] that for all s with Res large enough, all elements of the space $\mathcal{B}(\alpha^{s-\frac{1}{2}},\alpha^{\frac{1}{2}-s}\omega^{-1})$ have the form

$$f(g) = z(\alpha^{2s}\omega, g.\Phi)|\det g|^s ,$$

for a suitable Φ in $\mathcal{S}(F^2)$. If s and Φ are such that the function f vanishes, so does the function

$$h(g) = z(\alpha^{2-2s}\omega^{-1}, g.\hat{\Phi})|\det g|^{1-s}\omega^{-1}(\det g)$$

and therefore, by (14.7.3), the integral $\tilde{\Psi}(1-s,W_1,W_2,\hat{\Phi})$.

From this follows the first assertion of the next lemma.

Lemma 14.7.4: There is s_0 so that for Res $> s_0$ there is a unique trilinear form γ_s on the product

$$\mathcal{W}_1(\pi_1,\psi) \times \mathcal{W}_2(\pi_2,\psi) \times \mathcal{B}(\alpha^{s-\frac{1}{2}},\alpha^{\frac{1}{2}-s}\omega^{-1})$$

so that if f is the function defined by

$$f(g) = z(\alpha^{2s}\omega, g.\Phi)\,|\det g|^{s}$$

then

$$\gamma_s(W_1, W_2, f) = \tilde{\Psi}(1-s, W_1, W_2, \hat{\Phi})$$

for all W_i in $\mathcal{W}(\pi_i, \psi)$. Moreover, for all W_i and all f in $\mathcal{B}(\alpha^{s-\frac{1}{2}}, \alpha^{\frac{1}{2}-s}\omega^{-1})$ and all g

$$\gamma_s(\pi_1(g)W_1, \pi_2(g)W_2, \rho(g)f) = \gamma_s(W_1, W_2, f)\ .$$

(We denote by $\rho(g)f$ the right translate of f under g^{-1}).

If Res is small enough $\tilde{\Psi}(1-s, W_1, W_2, \hat{\Phi})$ is defined by an integral on $ZN\backslash G$ against the invariant measure. Then

$$\tilde{\Psi}(1-s, \pi_1(g)W_1, \pi_2(g)W_2, g.\hat{\Phi}) = |\det g|^{s-1}\omega(\det g)\tilde{\Psi}(1-s, W_1, W_2, \hat{\Phi})\ .$$

By analytic continuation this formula is true for almost all s. The definition of γ_s and some easy formal computations give the second assertion of the lemma.

As the functions of the Whittaker model are K-finite on the right, we see that for W_1 and W_2 fixed, the linear form

$$f \longmapsto \gamma_s(W_1, W_2, f)$$

is K-finite, i.e., belongs to the space of the representation contragredient to $\rho(\alpha^{s-\frac{1}{2}}, \alpha^{\frac{1}{2}-s}\omega^{-1})$ that is the representation $\rho(\alpha^{\frac{1}{2}-s}, \alpha^{s-\frac{1}{2}}\omega)$. Hence there is a function

$$g \longmapsto \delta_s(g, W_1, W_2)$$

which satisfies

$$\delta_s\left[\begin{pmatrix} a & x \\ 0 & b \end{pmatrix} g, W_1, W_2\right] = |a|^{1-s}|b|^{s-1}\omega(b)\delta_s(g, W_1, W_2)$$

and

$$\gamma_s(W_1,W_2,f) = \int_{P\backslash G} \delta_s(g,W_1,W_2)f(g)dg \quad .$$

Here we denote by P the group of triangular matrices and the "measure" which appears on the right-hand side is the invariant linear form on the space of all continuous functions on G_F , which, on the left, transform according to the module of P_F .

Clearly, the invariance property of γ_s implies, for δ_s , the invariance property

$$\delta_s(gx^{-1},\pi_1(x)W_1,\pi_2(x)W_2) = \delta_s(g,W_1,W_2) \quad .$$

In particular, if we set

$$\lambda_s(\varphi_1,\varphi_2) = \delta_s(e,W_1,W_2)$$

where

$$\varphi_i(a) = W_i\begin{pmatrix} a & 0 \\ 0 & 1 \end{pmatrix}$$

we see that λ_s is a bilinear form satisfying the assumptions of Lemma (14.4.2). So there is a function $c(s)$ of s , defined for Res large enough, so that

$$\lambda_s = c(s)\beta_s \quad .$$

Coming back to the definition of δ_s and γ_s we see that, for Res large enough,

$$\tilde{\Psi}(1-s,W_1,W_2,\hat{\Phi}) = c(s)\int_K f(k)\beta_s(\pi_1(k)\varphi_1,\pi_2(k)\varphi_2)dk \quad .$$

where

$$f(g) = z(\alpha^{2s}\omega,g.\Phi)\,|\det g|^s \quad .$$

Comparing with (14.7.2) we find that

$$\tilde{\Psi}(1-s,W_1,W_2,\hat{\Phi}) = c(s)\Psi(s,W_1,W_2,\Phi)$$

for Res large enough.

To conclude the proof of (14.7.3) it is enough to show that $c(s)$ is a rational function of q^{-s}. To do that it will suffice to apply the following lemma.

Lemma 14.7.5: One can find W_i, $i = 1,2$ and Φ in $\mathcal{S}(F^2)$ so that, for all s,

$$\Psi(s,W_1,W_2,\Phi) = 1 .$$

We first choose W_i so that

$$\varphi_i(a) = W_i\begin{pmatrix} a & 0 \\ 0 & 1 \end{pmatrix}$$

is the characteristic function of $R^\times$ in $F^\times$. There is $n \geq 1$ so that W_1 and W_2 are invariant under translations by the matrices

$$\begin{pmatrix} 1 & 0 \\ x & 1 \end{pmatrix} \text{ with } x \in \varpi^n R .$$

Then if K' is the subgroup of all k in K which have the form

$$k = \begin{pmatrix} a & b \\ c & d \end{pmatrix} \text{ with } c \equiv 0 \bmod \varpi^n R$$

we see that for k in K',

$$\int W_1\left[\begin{pmatrix} a & 0 \\ 0 & 1 \end{pmatrix}k\right] W_2\left[\begin{pmatrix} -a & 0 \\ 0 & 1 \end{pmatrix}k\right] |a|^{s-1} d^\times a = \text{meas } R^\times \omega(d) .$$

Now let Φ be the element of $\mathcal{S}(F^2)$ defined by

$$\begin{aligned} \Phi(x,y) &= \omega^{-1}(y) \text{ if } x \in \varpi^n R \text{ and } y \in R^\times \\ &= 0 \text{ otherwise.} \end{aligned}$$

Then, for $k \in K$,

$$Z(\alpha^{2s}\omega, k.\Phi) = \omega^{-1}(d) \quad \text{if} \quad k \in K' ,$$

$$= 0 \quad \text{otherwise.}$$

Hence we find

$$\Psi(s,W_1,W_2,\Phi) = (\text{meas } R^{\times})^2 .$$

The lemma follows.

As usual, we shall need more precise results.

Theorem 14.8: *There are Euler factors* $L(s,\pi)$ *and* $L(s,\tilde{\pi})$ *with the following properties.*

(1) *Set*

$$\Psi(s,W_1,W_2,\Phi) = L(s,\pi)\Xi(s,W_1,W_2,\Phi) ,$$

$$\tilde{\Psi}(s,W_1,W_2,\Phi) = L(s,\tilde{\pi})\tilde{\Xi}(s,W_1,W_2,\Phi) .$$

Then $\Xi(s,W_1,W_2,\Phi)$ *and* $\tilde{\Xi}(s,W_1,W_2,\Phi)$ *are polynomials in* q^{-s} *and* q^{s} .

(2) *One can choose families* W_1^i , W_2^i *and* Φ^i *so that*

$$\sum_i \Xi(s,W_1^i,W_2^i,\Phi^i) = 1 \quad (\text{resp. } \sum_i \tilde{\Xi}(s,W_1^i,W_2^i,\Phi^i) = 1) .$$

(3) *There is a function* $\epsilon(s,\pi,\psi)$ *which, as a function of* s , *has the form* cq^{-si} *so that*

$$\tilde{\Xi}(1-s,W_1,W_2,\hat{\Phi}) = \omega_2(-1)\epsilon(s,\pi,\psi)\Xi(s,W_1,W_2,\Phi) .$$

From the integral representation (14.5) follows that

(14.8.4) $$\Psi(s,\pi_1(g)W_1,\pi_2(g)W_2,g.\Phi) = |\det g|^{-s}\Psi(s,W_1,W_2,\Phi) .$$

Combining this with (14.7) we see that the sub-vector space of $\underline{C}(q^{-s})$ spanned by the $\Psi(s,W_1,W_2,\Phi)$ is, in fact, a fractional ideal of the ring $\underline{C}[q^{-s},q^{s}]$. Let $P_0.Q_0^{-1}$ be a generator of this ideal. We may assume that $Q_0(0) = 1$, $P_0(0) = 1$ or $P_0 = 0$ and P_0, Q_0 are relatively prime (if $P_0 \neq 0$) . From Lemma (14.7.5) follows that $P_0 = 1$. Then

$$L(s,\pi) = Q_0(q^{-s})^{-1}$$

is the unique Euler factor satisfying the conditions (1) and (2). There is, similarly, a unique Euler factor $L(s,\tilde{\pi})$ satisfying the two first conditions. Then there is a function of s which satisfy the functional equation (3). By (1) and (2) it is a polynomial in q^{-s} and q^{s} . But there is a similar factor $\epsilon(s,\tilde{\pi},\psi)$. Exchanging the roles of π and $\tilde{\pi}$ we find

$$\epsilon(s,\pi,\psi)\,\epsilon(1-s,\tilde{\pi},\psi) = 1 \quad .$$

Hence $\epsilon(s,\pi,\psi)$ must be a monomial. This concludes the proof of Theorem 14.8.

If we substitute to π the representation

$$\pi' = \pi_2 \times \pi_1$$

we easily see that

$$L(s,\pi) = L(s,\pi') \ , \ L(s,\tilde{\pi}) = L(s,\tilde{\pi}') \ , \ \epsilon(s,\pi,\psi) = \epsilon(s,\pi',\psi) \ .$$

Formal manipulations show that if the additive character ψ is replaced by the additive character

$$\psi'(x) = \psi(bx) \quad (b \in F^{\times})$$

the factor $\epsilon(s,\pi,\psi)$ is replaced by

$$\text{(14.8.5)} \qquad \epsilon(s,\pi,\psi') = \omega^2(b)\,|b|^{4(s-\frac{1}{2})}\,\epsilon(s,\pi,\psi) \ .$$

Finally, it is convenient to introduce the factor

$$\epsilon'(s,\pi,\psi) = \epsilon(s,\pi,\psi)L(1-s,\tilde{\pi})/L(s,\pi) \ .$$

Then the functional equation reads

(14.8.6) $$\tilde{\Psi}(1-s,W_1,W_2,\hat{\Phi}) = \omega_2(-1)\Psi(s,W_1,W_2,\Phi).\epsilon'(s,\pi,\psi) .$$

<u>Remark 14.9</u>: Suppose that $\pi_1 = \sigma(\mu_1,\nu_1)$ where $\mu_1.\nu_1^{-1} = \alpha$. Then $\mathcal{W}(\pi_1,\psi)$ is a subspace of codimension one in $\mathcal{W}(\mu_1,\nu_1,\psi)$. For W_1 in that larger space and W_2 in $\mathcal{W}(\pi_2,\psi)$ the integrals $\Psi(s,W_1,W_2,\Phi)$ and $\tilde{\Psi}(s,W_1,W_2,\Phi)$ can be defined. The above results, suitably modified, can be applied to them. In particular, they are convergent for Res large enough, are rational functions of q^{-s} and satisfy the functional equation (14.8.6).

If π_1 and π_2 are special, a similar remark can be made.

§15. Explicit computations

The purpose of §15 is to prove the following theorem.

Theorem 15.1: Let π_i , $i = 1,2$ be two admissible irreducible representations of $GL(2,F)$. Let π be $\pi_1 \times \pi_2$. Assume that neither π_1 nor π_2 is one dimensional.

(1) If $\pi_2 = \pi(\mu_2,\nu_2)$ then

$$L(s,\pi) = L(s,\pi_1 \otimes \mu_2)L(s,\pi_1 \otimes \nu_2) \quad ,$$
$$L(s,\tilde{\pi}) = L(s,\tilde{\pi}_1 \otimes \mu_2^{-1})L(s,\tilde{\pi}_1 \otimes \nu_2^{-1}) \quad ,$$
$$\epsilon(s,\pi,\psi) = \epsilon(s,\pi_1 \otimes \mu_2,\psi)\epsilon(s,\pi_1 \otimes \nu_2,\psi) \ ;$$

(2) If $\pi_2 = \sigma(\mu_2,\nu_2)$ with $\mu_2.\nu_2^{-1} = \alpha_F$ then

$$L(s,\pi) = L(s,\pi_1 \otimes \mu_2) \quad ,$$
$$L(s,\tilde{\pi}) = L(s,\tilde{\pi}_1 \otimes \nu_2^{-1}) \ ,$$
$$\epsilon(s,\pi,\psi) = \epsilon(s,\pi_1\otimes\mu_2,\psi)\epsilon(s,\pi_1\otimes\nu_2,\psi)L(s,\pi_1\otimes\nu_2)^{-1}L(1-s,\tilde{\pi}_1\otimes\mu_2^{-1}) \ .$$

The theorem implies the following result.

Proposition 15.2: With the notations and assumptions of Theorem 15.1 for W_i in $W(\pi_i,\psi)$, we have

$$\tilde{\Psi}(1-s,W_1,W_2,\hat{\Phi}) = \epsilon'(s,\pi_1\otimes\mu_2,\psi)\epsilon'(s,\pi_1\otimes\nu_2,\psi)\omega_2(-1)\Psi(s,W_1,W_2,\Phi) \ .$$

In fact, we shall deduct 15.1 from 15.2. For the time being, we take 15.2 for granted and prove the theorem. All we have to show is that $L(s,\pi)$ has the required values. We shall use a series of lemmas.

Lemma 15.3: Assume $\pi_2 = \sigma(\mu_2,\nu_2)$ where $\mu_2.\nu_2^{-1} = \alpha$.

(1) If φ_i belongs to $\mathcal{K}(\pi_i,\psi)$, $i = 1,2$,

$$\beta_s(\varphi_1,\varphi_2) = L(s,\pi_1 \otimes \mu_2)P(s)$$

<u>where</u> P <u>belongs to</u> $\underline{C}[q^{-s},q^{s}]$.

(2) <u>There are families</u> φ_1^j <u>in</u> $\mathcal{K}(\pi_1,\psi)$ <u>and</u> φ_2^j <u>in</u> $\mathcal{K}(\pi_2,\psi)$ <u>so that</u>

$$\sum_j \beta_s(\varphi_1^j,\varphi_2^j) = L(s,\pi_1 \otimes \mu_2) \quad .$$

We recall that the subvector space of $\underline{C}(q^{-s})$ spanned by the integrals

$$\int \varphi_1(a)\mu_2(a)|a|^{s-\frac{1}{2}} d^{\times}a \ , \ \varphi_1 \in \mathcal{K}(\pi_1,\mu) \ ,$$

is in fact a fractional ideal of the ring $\underline{C}[q^{-s},q^{s}]$ of which the Euler factor $L(s,\pi_1 \otimes \mu_2)$ is a generator. On the other hand, by Lemma 14.2, we know that all elements of $\mathcal{K}(\pi_2,\psi)$ have the form

$$\varphi_2(a) = f(a)\mu_2(a)|a|^{\frac{1}{2}}$$

where f belongs to $\mathcal{S}(F)$. From the same lemma follows that the function

$$a \longmapsto \varphi_1(a)f(a) \ , \ \varphi_1 \in \mathcal{K}(\pi_1,\psi) \ ,$$

is also in $\mathcal{K}(\pi_1,\psi)$. Therefore, we find that

$$\beta_s(\varphi_1,\varphi_2) = \int \varphi_1(a)f(a)\mu_2(a)|a|^{s-\frac{1}{2}} d^{\times}a = L(s,\pi_1\otimes\mu_2)P(s)$$

where P belongs to $\underline{C}[q^{-s},q^{s}]$. Hence the first assertion of the lemma.

If φ_1 is given in $\mathcal{K}(\pi_1,\psi)$ its support is contained in some set $\varpi^{i}R - \{0\}$. (Cf. [I] §2). If f is the characteristic function of $\varpi^{i}R$ in F , it belongs to $\mathcal{S}(F)$ and satisfies

$$\varphi_1(a)f(a) = \varphi_1(a) \quad \text{for all} \quad a \in F^{\times} \ .$$

Again, by Lemma 1.4 the function φ_2 defined by

$$\varphi_2(a) = f(a)\mu_2(a)|a|^{\frac{1}{2}}$$

belongs to $\mathcal{K}(\pi_2,\psi)$. We find

$$\beta_s(\varphi_1,\varphi_2) = \int \varphi_1(a)|a|^{s-\frac{1}{2}}\mu_2(a)d^{\times}a \quad .$$

The second assertion of the lemma follows.

<u>Lemma 15.4</u>: <u>Assume</u> $\pi_2 = \pi(\mu_2,\nu_2)$ <u>where</u> $\mu_2.\nu_2^{-1}$ <u>is different from</u> α, α^{-1}.

(1) <u>For</u> φ_i <u>in</u> $\mathcal{K}(\pi_i,\psi)$

$$\beta_s(\varphi_1,\varphi_2) = L(s,\pi_1\otimes\mu_2)L(s,\pi_1\otimes\nu_2)P(s)$$

<u>where</u> P <u>belongs to</u> $\underline{C}[q^{-s},q^{s}]$.

(2) <u>If</u> $L(s,\pi_1\otimes\mu_2)^{-1}$ <u>and</u> $L(s,\pi_1\otimes\nu_2)^{-1}$ <u>are relatively prime elements of</u> $\underline{C}[q^{-s}]$ <u>there are families</u> φ_1^j <u>and</u> φ_2^j <u>so that</u>

$$\sum_j \beta_s(\varphi_1^j,\varphi_2^j) = L(s,\pi_1\otimes\mu_2)L(s,\pi_1\otimes\nu_2) \quad .$$

If the representation π_1 is absolutely cuspidal we know that

$$L(s,\pi_1\otimes\mu_2) = L(s,\pi_1\otimes\nu_2) = 1$$

and

$$\mathcal{K}(\pi_1,\psi) = \mathcal{S}(F^{\times}) \quad .$$

Then both assertions are obvious.

If $\pi_1 = \sigma(\mu_1,\nu_1)$ with $\mu_1.\nu_1^{-1} = \alpha$ then

$$L(s,\pi_1\otimes\mu_2)L(s,\pi_1\otimes\nu_2) = L(s,\mu_1\mu_2)L(s,\mu_1\nu_2) = L(s,\pi_2\otimes\mu_1) \quad .$$

So in that case we just exchange the roles of the representations π_1 and π_2 and reduce Lemma 15.4 to Lemma 15.3.

Finally, assume $\pi_1 = \pi(\mu_1,\nu_1)$. Then we note that

$$L(s,\pi_1 \otimes \mu_2)L(s,\pi_1 \otimes \nu_2) = L(s,\pi_2 \otimes \mu_1)L(s,\pi_2 \otimes \nu_1) .$$

Therefore we may at will exchange the roles of π_1 and π_2.

Suppose first that μ_2 is different from ν_2. Then every element of $K(\pi_2,\psi)$ has the form

$$\varphi_2(a) = f(a)\mu_2(a)|a|^{\frac{1}{2}} + h(a)\nu_2(a)|a|^{\frac{1}{2}}$$

where f and h belong to $\mathcal{S}(F)$. With these notations we find

$$\beta_s(\varphi_1,\varphi_2) = \int f(a)\varphi_1(a)\mu_2(a)|a|^{s-\frac{1}{2}}d^\times a + \int h(a)\varphi_1(a)\nu_2(a)|a|^{s-\frac{1}{2}}d^\times a .$$

As $f\varphi_1$ and $h\varphi_1$ belong to $\mathcal{K}(\pi_1,\psi)$ we see that the right-hand side has the form

$$L(s,\pi_1 \otimes \mu_2)P(s) + L(s,\pi_1 \otimes \nu_2)Q(s) = L(s,\pi_1 \otimes \mu_2)\, L(s,\pi_1 \otimes \nu_2)R(s)$$

where P, Q and R belong to $\underline{C}[q^{-s},q^s]$. So the first assertion of the lemma follows.

If $\mu_2 = \nu_2$ but $\mu_1 \neq \nu_1$ we may exchange the roles of π_1 and π_2 and reduce ourselves to the previous case.

Finally, if $\mu_1 = \nu_1$ and $\mu_2 = \nu_2$ we see that $\beta_s(\varphi_1,\varphi_2)$ is a sum of terms of the form

$$\int f(a)\mu_1\mu_2(a)|a|^s v^i(a)d^\times a$$

where $i = 0,1,2$ and $f \in \mathcal{S}(F)$. If $\mu_1\mu_2$ is ramified this is a polynomial in q^{-s} and q^s. If $\mu_1\mu_2$ is unramified this has the form

$$L(s,\mu_1\mu_2)^3\, P(s)$$

where P belongs to $\underline{C}[q^{-s},q^s]$. (Cf. Lemma 14.1). Hence the first assertion of the lemma follows.

Assume now that $L(s,\pi_1 \otimes \mu_2)^{-1}$ and $L(s,\pi_1 \otimes \nu_2)^{-1}$ are relatively prime. There are P and Q in $\underline{C}[q^{-s}]$ so that

$$L(s,\pi_1 \otimes \mu_2)L(s,\pi_1 \otimes \nu_2) = PL(s,\pi_1 \otimes \mu_2) + QL(s,\pi_1 \otimes \nu_2) .$$

There are also φ_1' and φ_2'' in $\mathcal{K}(\pi_1,\psi)$ so that

$$\int \varphi_1'(a)\mu_2(a)|a|^{s-\frac{1}{2}} d^{\times}a = PL(s,\pi_1 \otimes \mu_2) ,$$

$$\int \varphi_1''(a)\mu_2(a)|a|^{s-\frac{1}{2}} d^{\times}a = QL(s,\pi_1 \otimes \nu_2) ,$$

and f in $\mathcal{S}(F)$ so that

$$\varphi_1' f = \varphi_1' , \quad \varphi_2'' f = \varphi_2'' .$$

If we set now

$$\varphi_2'(a) = f(a)\mu_2(a)|a|^{\frac{1}{2}} , \quad \varphi_2''(a) = f(a)\nu_2(a)|a|^{\frac{1}{2}}$$

we find

$$\beta_s(\varphi_1',\varphi_2') + \beta_s(\varphi_1'',\varphi_2'') = L(s,\pi_1 \otimes \mu_2)L(s,\pi_1 \otimes \nu_2) .$$

So the lemma is completely proved.

<u>Lemma 15.5</u>: <u>Assume that</u> $\pi_2 = \pi(\mu_2,\nu_2)$. <u>Then the quotients</u>

$$\Psi(s,W_1,W_2,\Phi)/L(s,\pi_1 \otimes \mu_2)L(s,\pi_1 \otimes \nu_2)$$

<u>belong to</u> $\underline{C}[q^{-s},q^{s}]$.

By the previous lemma and Lemma 14.7.2 we find that such a quotient has the form $L(2s,\omega)P(q^{-s},q^{s})$ where P is a polynomial, ω is the product $\omega_1\omega_2$ and

$$\pi_i(a) = \omega_i(a) , \ i = 1,2, \quad a \in F^{\times} .$$

If ω is ramified $L(2s,\omega) = 1$ and the assertion of (15.5) is proved. If ω is unramified we may as well assume that $\omega = 1$. Then if

$\Phi(0,0) = 0$, for each $k \in K$, the expression

$$z(\alpha^{2s}\omega, k.\Phi)$$

belongs to $\underline{C}[q^{-s}, q^{s}]$; and again (15.5) is proved. If $\Phi(0,0)$ does not vanish, by linearity, we may assume that Φ is the characteristic function of R^2 in F^2 . There is also no harm in assuming that ψ has order 0 . Then $\Phi = \hat{\Phi}$ and

$$\tilde{\Psi}(s, W_1, W_2, \hat{\Phi}) = \Psi(s, W_1, W_2, \Phi)$$
$$= P(X, X^{-1})(1-X^2)^{-1} L(s, \pi_1 \otimes \mu_2) L(s, \pi_1 \otimes \nu_2)$$

where $X = q^{-s}$ and P is a polynomial. As we assume

$$\omega_1 \mu_2 \nu_2 = \omega = 1$$

we find

$$L(s, \pi_1 \otimes \mu_2) L(s, \pi_1 \otimes \nu_2) = L(s, \tilde{\pi}_1 \otimes \nu_2^{-1}) L(s, \tilde{\pi}_1 \otimes \mu_2^{-1}) .$$

Therefore the functional equation of (15.2) (that we take for granted) reads (with $X = q^{-s}$)

$$P(q^{-1}X^{-1}, qX)(1-q^{-2}X^{-2})^{-1} = cX^{i} P(X, X^{-1})(1-X^2)^{-1} .$$

This obviously implies that

$$P(X, X^{-1})(1-X^2)^{-1}$$

belongs to $\underline{C}[X, X^{-1}]$ which is precisely our assertion.

<u>Lemma 15.6</u>: <u>Assume</u> $\pi_2 = \sigma(\mu_2, \nu_2)$ <u>with</u> $\mu_2 \nu_2^{-1} = \alpha$. <u>Then the quotients</u>

$$\Psi(s, W_1, W_2, \Phi)/L(s, \pi_1 \otimes \mu_2)$$

<u>belong to</u> $\underline{C}[q^{-s}, q^{2}]$.

The proof is similar.

We now conclude the proof of (15.1) (when (15.2) is granted).

Lemma 15.7: If $\pi_2 = \sigma(\mu_2,\nu_2)$ with $\mu_2.\nu_2^{-1} = \alpha$, then

$$L(s,\pi) = L(s,\pi_1 \otimes \mu_2) .$$

From (15.3) we know the existence of some families W_i^j in $\mathcal{W}(\pi_i,\psi)$ so that

$$\sum_j \int W_1^j\begin{pmatrix} a & 0 \\ 0 & 1 \end{pmatrix} W_2^j\begin{pmatrix} -a & 0 \\ 0 & 1 \end{pmatrix} |a|^{s-1}\, d^\times a = L(s,\pi_1 \otimes \mu_2) .$$

There is $n > 0$ so that W_1^j and W_2^j are invariant under left translations by the matrices:

$$\begin{pmatrix} 1 & 0 \\ x & 1 \end{pmatrix} \quad \text{where} \quad x \equiv 0 \bmod \varpi^n R .$$

Let K' be the subgroup of K whose elements have the form

$$\begin{pmatrix} a & b \\ c & d \end{pmatrix} \quad \text{with} \quad c \equiv 0 \bmod \varpi^n R .$$

Then for k in K' we find

$$\sum_j \int W_1^j\left[\begin{pmatrix} a & 0 \\ 0 & 1 \end{pmatrix} k\right] W_2^j\left[\begin{pmatrix} -a & 0 \\ 0 & 1 \end{pmatrix} k\right] |a|^{s-1} d^\times a = L(s,\pi_1 \otimes \mu_2)\omega(d) .$$

If we choose Φ in $\mathcal{S}(F^2)$ so that, for k in K

$$f(k) = z(\alpha^{2s}\omega, k.\Phi) = \omega^{-1}(d) \quad \text{if} \quad k \in K' ,$$

0 otherwise, (cf. (14.7.5)), we get

$$\sum_j \Psi(s,W_1^j,W_2^j,\Phi) = L(s,\pi_1 \otimes \mu_2)\,\text{meas}(K') .$$

Hence the lemma.

Lemma 15.8: If $\pi_2 = \pi(\mu_2,\nu_2)$ then

$$L(s,\pi) = L(s,\pi_1 \otimes \mu_2) L(s,\pi_1 \otimes \nu_2) .$$

If the polynomials in q^{-s}

$$L(s,\pi_1 \otimes \mu_2)^{-1} \quad \text{and} \quad L(s,\pi_1 \otimes \nu_2)^{-1}$$

are relatively prime, we use (15.4.2) and argue as in (15.7). So we may assume that these polynomials are not relatively prime. In that case

$$\pi_1 = \sigma(\mu_1,\nu_1) \quad \text{or} \quad \pi_1 = \pi(\mu_1,\nu_1) \ .$$

But if π_1 is special we may exchange the roles of π_1 and π_2 and apply (15.7). So we may assume that $\pi_1 = \pi(\mu_1,\nu_1)$. Then

(15.8.1) $\mu_1.\nu_1^{-1} \neq \alpha,\alpha^{-1}$, $\mu_2.\nu_2^{-1} \neq \alpha,\alpha^{-1}$.

Since

$$L(s,\pi_1 \otimes \mu_2) = L(s,\mu_1\mu_2)L(s,\nu_1\mu_2)$$

$$L(s,\pi_1 \otimes \nu_2) = L(s,\mu_1\nu_2)L(s,\nu_1\nu_2)$$

we see that at least one of the following relations is true:

$$\mu_2 = \nu_2 \qquad \text{is unramified}$$

(15.8.2) $$\mu_1\mu_2 = \nu_1\nu_2 \qquad \text{is unramified}$$

$$\mu_1\mu_2 = \nu_1\nu_2 \qquad \text{is unramified.}$$

On the other hand the functional equation (15.2) implies that

(15.8.3) $L(1-s,\tilde{\pi})/L(s,\pi) =$

$$\frac{L(1-s,\mu_1^{-1}\mu_2^{-1})L(1-s,\nu_1^{-1}\mu_2^{-1})L(1-s,\mu_1^{-1}\nu_2^{-1})L(1-s,\nu_1^{-1}\nu_2^{-1})}{L(s,\mu_1\mu_2)L(s,\nu_1\mu_2)L(s,\mu_1\nu_2)L(s,\nu_1\nu_2)}$$

up to multiplication by a factor of the form cq^{-ms} . Moreover, by Lemma 15.6 the quotients

$$L(s,\pi)/L(s,\mu_1\mu_2)L(s,\nu_1\mu_2)L(s,\mu_1\nu_2)L(s,\nu_1\nu_2)$$

$$L(s,\tilde{\pi})/L(s,\mu_1^{-1}\mu_2^{-1})L(s,\nu_1^{-1}\mu_2^{-1})L(s,\mu_1^{-1}\nu_2^{-1})L(s,\nu_1^{-1}\nu_2^{-1})$$

are polynomial in q^{-s} .

So it is clear that, if the rational fraction on the right-hand side of (15.8.3) is irreducible, Lemma 15.8 is true. Assume now that this rational fraction is not irreducible. By exchanging the roles of the μ_i's and ν_i's we may assume that $\mu_1\mu_2$ is unramified and satisfy one of the following relations:

$$\mu_1\mu_2\nu_1^{-1}\mu_2^{-1} = \alpha^{-1} \ , \ \mu_1\mu_2\mu_1^{-1}\nu_2^{-1} = \alpha^{-1} \ , \ \mu_1\mu_2\nu_1^{-1}\nu_2^{-1} = \alpha^{-1} \ .$$

Taking (15.8.1) in account we see that actually only the last case is possible. If we compare to (15.8.2) we see that

$$\nu_1\mu_2 = \mu_1\nu_2 \quad \text{is unramified.}$$

So the four products $\mu_1\mu_2$, $\mu_1\nu_2$, $\nu_1\mu_2$, $\nu_1\nu_2$ are unramified. Replacing (π_1,π_2) by $(\pi_1 \otimes \mu_2, \pi_2 \otimes \mu_2^{-1})$ we see that we may assume that the four quasi-characters μ_i and ν_i are unramified. This case will be taken up in Proposition 15.9 ("unramified situation").

So, except in the "unramified situation", Theorem 15.1 is a corollary of Proposition 15.2.

We now prove Proposition 15.2. We start by the "unramified" situation. Therefore we take for the invariant measure on $ZN\backslash G$ the measure defined by

$$\int_{ZN\backslash G} f(g)dg = \int_{F^\times \times K} f\left[\begin{pmatrix} a & 0 \\ 0 & 1 \end{pmatrix} k\right] |a|^{-1} \, d^\times a\, dk$$

where dk is the normalized Haar measure on K and $d^\times a$ the Haar measure on $F^\times$ for which the measure of $R^\times$ is one. We assume the additive character ψ to be of order zero. We suppose that both

π_1 and π_2 contain the trivial representation of K. So there are two pairs (μ_1,ν_1) and (μ_2,ν_2) of unramified quasi-characters of $F^\times$ such that

$$\pi_i = \pi(\mu_i,\nu_i) , \quad i = 1,2 .$$

We define also the quasi-characters

$$\mu_i'(x) = |\mu_i(x)| , \quad \nu_i'(x) = |\nu_i(x)| , \quad \text{(ordinary absolute value)}$$

$$\omega' = \mu_1'\mu_2'\nu_1'\nu_2' \quad .$$

Proposition 15.9: (1) Let Φ be the characteristic function of R^2 in F^2 and, for $i = 1,2$, W_i the unique element of $W(\pi_i,\psi)$ which is right invariant under K and takes the value one on K. Then

$$\Psi(s,W_1,W_2,\Phi) = L(s,\mu_1\mu_2)L(s,\mu_1\nu_2)L(s,\nu_1\mu_2)L(s,\nu_1\nu_2) .$$

(2) We have

$$L(s,\pi) = L(s,\mu_1\mu_2)L(s,\mu_1\nu_2)L(s,\nu_1\mu_2)L(s,\nu_1\nu_2) ,$$

$$L(s,\tilde{\pi}) = L(s,\mu_1^{-1}\mu_2^{-1})L(s,\mu_1^{-1}\nu_2^{-1})L(s,\nu_1^{-1}\mu_2^{-1})L(s,\nu_1^{-1}\nu_2^{-1})$$

$$\epsilon(s,\pi,\psi) = \epsilon(s,\tilde{\pi},\psi) = 1 .$$

(3) If s is real and so large that all the power series (in q^{-s})

$$L(s,\omega') , \ L(s,\mu_1'\mu_2') , \ L(s,\mu_1'\nu_2') , \ L(s,\nu_1'\mu_2') , \ L(s,\nu_1'\nu_2') ,$$

are convergent, we have the majorization

$$\int_{ZN\backslash G} |W_1(g)W_2(\eta g)z(\alpha^{2s}\omega,g.\Phi)|\,|\det g|^s\,dg$$

$$\leq L(s,\mu_1'\mu_2')L(s,\mu_1'\nu_2')L(s,\nu_1'\mu_2')L(s,\nu_1'\nu_2') .$$

Let φ_i be the element of $K(\pi_i,\psi)$ which corresponds to W_i.

Then $\varphi_i(\epsilon a) = \varphi_i(a)$ for ϵ in $R^\times$. In particular, we find

$$\Psi(s,W_1,W_2,\Phi) = L(2s,\omega)\int\varphi_1(a)\varphi_2(a)\,|a|^{s-1}\,d^\times a$$

$$= L(2s,\omega)\sum \varphi_1(\varpi^n)\varphi_2(\varpi^n)\;q^{-n(s-1)} \quad .$$

On the other hand, we have for $i = 1,2$

$$\int\varphi_i(a)\,|a|^{s-\frac{1}{2}}\,d^\times a = \sum \varphi_i(\varpi^n)\;q^{-n(s-\frac{1}{2})} = L(s,\mu_i)L(s,\nu_i) \;.$$

To prove the first assertion of the lemma we have only to apply the following lemma:

Lemma 15.9.4: Consider the formal series

$$\sum a_n' X^n = \frac{1}{(1-a'X)(1-b'X)} \;,\quad \sum a_n'' X^n = \frac{1}{(1-a''X)(1-b''X)} \quad .$$

Then

$$\sum a_n' a_n'' X^n = \frac{1-a'a''b'b''X^2}{(1-a'a''X)(1-a'b''X)(1-b'a''X)(1-b'b''X)} \quad .$$

We leave the proof to the reader as a refreshing exercise. (Hint: decompose the rational fractions into their simple elements).

The quasi-character

$$\omega = \omega_1\omega_2 = \mu_1\nu_1\mu_2\nu_2$$

is unramified. It has the form

$$\omega = \alpha^s \;.$$

In particular, with the notations of the proposition $\Phi = \hat{\Phi}$ and

$$\tilde{\Psi}(1-s,W_1,W_2,\hat{\Phi}) = \Psi(1-s-\sigma,W_1,W_2,\Phi) \quad .$$

It follows that the functional equation (15.2) is true for that particular choice of W_1 , W_2 and Φ . By Proposition 14.7 it is true for all choices. So (15.2) and therefore (15.1) is true for the pair (π_1,π_2) . The second assertion of (15.9) follows from (15.5).

For the last assertion we take s real and so large that all the indicated power series converge. Then we note that

$$|z(\alpha^{2s}\omega, g.\Phi)| \leq z(\alpha^{2s}\omega', g.\Phi) .$$

On the other hand, the function φ_i has the integral representation

$$\varphi_i(a) = \mu_1(a)|a|^{\frac{1}{2}} \int \Phi(at, t^{-1})\mu_1\mu_2^{-1}(t)d^\times t .$$

So it is absolutely bounded by the function φ_i' defined by

$$\varphi_i'(a) = \mu_1'(a)|a|^{\frac{1}{2}} \int \Phi(at, t^{-1})\mu_1'\mu_2'^{-1}(t)d^\times t .$$

The integral of the function $\varphi_1'\varphi_2'\alpha^{s-1}$ on $F^\times$ is computed as the integral of the function $\varphi_1\varphi_2\alpha^{s-1}$ with μ_i' (resp. ν_i') replacing μ_i (resp. ν_i) .

Keeping in mind those remarks we find

$$\int_{ZN\backslash G} |W_1(g)W_2(\eta g)z(\alpha^{2s}\omega, g.\Phi)|\,|\det g|^s\, dg$$

$$\leq \int |W_1(g)W_2(\eta g)|\,|z(\alpha^{2s}\omega', g.\Phi)|\,|\det g|^s\, dg = \int |\varphi_1(a)\varphi_2(a)|\,|a|^{s-1}d^\times a L(2s,\omega')$$

$$\leq \int \varphi_1'(a)\varphi_2'(a)|a|^{s-1}\, d^\times a L(2s,\omega') = L(s,\mu_1'\mu_2')L(s,\mu_1'\nu_2')L(s,\nu_1'\mu_2')L(s,\nu_1'\nu_2').$$

Hence Proposition 15.9 is completely proved, and Theorem 15.1 is now a consequence of (15.2).

In general, to prove Proposition 15.2 we have only to check the functional equation for one triple (W_1, W_2, Φ) , this triple being such that $\Psi(s, W_1, W_2, \Phi)$ is not identically zero. We may also assume ψ to be of order zero. To simplify notations, we write $\epsilon(s,\pi)$, $\epsilon(s,\mu)$ and so on for $\epsilon(s,\pi,\psi)$, $\epsilon(s,\mu,\psi)$ .... . We further assume that π_2 has the form $\pi(\mu_2,\nu_2)$ and π_1 is absolutely cuspidal or has the form $\pi(\mu_1,\nu_1)$.

In fact, there is no harm in doing that. For if π_2, for instance, has the form $\sigma(\mu_2,\nu_2)$ then the functional equation (15.2) is valid for W_1 in $\mathcal{W}(\pi_1,\mu)$ and W_2 in $\mathcal{W}(\mu_2,\nu_2,\psi)$ which contains $\mathcal{W}(\pi_2,\psi)$. So with trivial modifications the forthcoming computations apply to that case. A similar remark can be made if both π_1 and π_2 are special.

We have to distinguish between the following cases

$$L(s,\pi_1 \otimes \mu_2)L(s,\pi_1 \otimes \nu_2) \neq 1$$

$$L(s,\pi_1 \otimes \mu_2)L(s,\pi_1 \otimes \nu_2) = 1 \quad .$$

In the first case the representation π_1 has the form $\pi(\mu_1,\nu_1)$ and at least one of the quasi-characters

$$\mu_1\mu_2 \; , \; \mu_1\nu_2 \; , \; \nu_1\mu_2 \; , \; \nu_1\nu_2$$

is unramified. There is no harm in assuming that it is $\nu_1\nu_2$. We may also replace the pair (π_1,π_2) by the pair $(\pi_1 \otimes \nu_2, \pi_2 \otimes \nu_2^{-1})$ and consequently assume that ν_1 and ν_2 are unramified. We are then led to distinguish the following cases; where $O(.)$ denotes the order,

(15.10) $\qquad \pi_1 = \pi(\mu_1,\nu_1)$ and $O(\mu_1) = O(\nu_1) = O(\mu_2) = O(\nu_2) = 0$.

(15.11) $\qquad \pi_1 = \pi(\mu_1,\nu_1)$ and $O(\mu_1) > 0$, $O(\nu_1) = O(\mu_2) = O(\nu_2) = 0$.

(15.12) $\qquad \pi_1 = \pi(\mu_1,\nu_1)$ and $O(\nu) = O(\nu_2) = 0$, $O(\mu_1\mu_2) = O(\mu_1) \geq O(\mu_2) > 0$.

(15.13) $\qquad \pi_1 = \pi(\mu_1,\nu_1)$ and $O(\nu_1) = O(\nu_2) = 0$, $O(\mu_1) = O(\mu_2) \geq O(\mu_1\mu_2) > 0$.

(15.14) $\qquad \pi_1 = \pi(\mu_1,\nu_1)$ and $O(\nu_1) = O(\nu_2) = 0$, $O(\mu_1) = O(\mu_2) > O(\mu_1\mu_2) = 0$.

(15.15) π_1 absolutely cuspidal or, $\pi_1 = \pi(\mu_1, \nu_1)$ and all the products

$$\mu_1\mu_2 \, , \, \mu_1\nu_2 \, , \, \nu_1\mu_2 \, , \, \nu_1\nu_2$$

are ramified.

The case (15.10) has already been taken care of in (15.9). The cases (15.11) to (15.14) will now be taken up. In those cases, we shall make use of some simple facts that we now state. For $\Phi(x,y) = \Phi_1(x)\Phi_2(y)$ and Res large enough $\omega(-1)\Psi(s,W_1,W_2,\Phi) =$

(15.10.1) $$\int W_1\left[\begin{pmatrix} a & 0 \\ 0 & 1\end{pmatrix} w \begin{pmatrix} 1 & x \\ 0 & 1\end{pmatrix}\right] W_2\left[\begin{pmatrix} -a & 0 \\ 0 & 1\end{pmatrix} w \begin{pmatrix} 1 & x \\ 0 & 1\end{pmatrix}\right]$$
$$|a|^{s-1}\, d^\times a dx \int \Phi_1(t)\Phi_2(tx)|t|^{2s}\omega(t)d^\times t \quad .$$

Similarly for Res small enough

$$\omega(-1)\tilde{\Psi}(1-s,W_1,W_2,\hat{\Phi}) =$$

(15.10.2)
$$\int W_1\left[\begin{pmatrix} a & 0 \\ 0 & 1\end{pmatrix} w \begin{pmatrix} 1 & x \\ 0 & 1\end{pmatrix} w\right] W_2\left[\begin{pmatrix} -a & 0 \\ 0 & 1\end{pmatrix} w \begin{pmatrix} 1 & x \\ 0 & 1\end{pmatrix} w\right]$$
$$\omega^{-1}(a)\, |a|^{-s}\, d^\times a dx \int \hat{\Phi}_1(t)\hat{\Phi}_2(tx)\, |t|^{2-2s}\omega^{-1}(t)d^\times t \quad .$$

In these formulas the integration is for x in F, t in $F^\times$ and a in $F^\times$. The additive Haar measure dx is so chosen that the measure of R is one. The multiplicative Haar measure $d^\times t$ or $d^\times a$ is so chosen that the measure of $R^\times$ is one. Both formulas are obvious if we remember that

$$\int_{ZN\backslash G} f(g)dg = \int f\left[\begin{pmatrix} a & 0 \\ 0 & 1\end{pmatrix} w \begin{pmatrix} 1 & x \\ 0 & 1\end{pmatrix}\right] |a|^{-1}\, d^\times a dx$$

where, as usual,

$$w = \begin{pmatrix} 0 & 1 \\ -1 & 0\end{pmatrix} ,$$

defines an invariant measure on $ZN\backslash G$.

It is also convenient to record as a lemma the following facts.

Lemma 15.10.3: Let μ be an unramified quasi-character and ν a quasi-character of order $u > 0$. Then

$$\epsilon(s,\mu) = 1, \ \epsilon(s,\nu) = q^{-us}\epsilon(0,\nu), \ \epsilon(s,\mu\nu) = \mu(\varpi^u)\epsilon(s,\nu) ,$$

and

$$\epsilon(s,\nu)\epsilon(1-s,\nu^{-1}) = \nu(-1) .$$

All assertions are simple consequences of the definitions. Finally we state without proof the following simple lemmas.

Lemma 15.10.4: Consider the formal series

$$\sum' a_n X^n = X^{-u}(1-aX)^{-1} , \sum' b_n X^n = X^{-v}(1-bX)^{-1}$$

where $u \geq v$. Then

$$\sum a_n b_n X^n = a^{u-v} X^{-v}(1-abX)^{-1} .$$

Lemma 15.10.5: Consider the formal series

$$\sum a_n X^n = (1-aX)^{-1}(1-bX)^{-1} , \sum' b_n X^n = X^{-u}(1-cX)^{-1} ,$$

where $u \geq 0$. Then

$$\sum a_n b_n X^n = c^u(1-acX)^{-1}(1-bcX)^{-1} .$$

In fact, both can be used as steps in the proof of (15.9.4).

We are now ready to take up the cases (15.11) to (15.14).

Case (15.11): We assume that $\pi_1 = \pi(\mu_1,\nu_1)$ and $\pi_2 = \pi(\mu_2,\nu_2)$ with

$$u = O(\mu_1) = O(\omega) > 0 , \ O(\nu_1) = O(\mu_2) = O(\nu_2) = 0 .$$

We denote by Φ_0 the characteristic function of R in F . For a ramified character like ω , which is of order $u > 0$, we denote by

Φ_ω the function which is equal to $\omega^{-1}(x)$ if x is in $R^\times$ and zero otherwise. Its Fourier transform has then a support contained in $\varpi^{-u}R^\times$. Moreover

$$\int \Phi_\omega(t)\,|t|^s\, d^\times t = 1 \;,\; \int \hat{\Phi}_\omega(t)\,|t|^{1-s}\, d^\times t = \epsilon(s,\omega) \;.$$

We choose W_1, W_2 and Φ in the following manner. First,

$$\Phi(x,y) = \Phi_\omega(x)\Phi_0(y) \;.$$

As for W_i it corresponds to the element φ_i of $\mathcal{K}(\pi_i,\psi)$ which itself is defined by the following conditions.

For φ_1:
$$\varphi_1(a\epsilon) = \varphi_1(a)\omega_1(\epsilon) \quad \text{if} \quad \epsilon \in R^\times \;,$$
$$\int \varphi_1(a)\,|a|^{s-\frac{1}{2}}\omega_1^{-1}(a)\,d^\times a = L(s,\nu_1^{-1}),$$
$$\int \pi_1(w)\varphi_1(a)\,|a|^{s-\frac{1}{2}}\, d^\times a = L(s,\nu_1)\,\epsilon(1-s,\mu_1^{-1})$$
$$= L(s,\nu_1)\,\epsilon(1,\mu_1^{-1})\,q^{us} \;.$$

For φ_2:
$$\varphi_2(a\epsilon) = \varphi_2(a) \quad \text{if} \quad \epsilon \in R^\times \;,$$
$$\int \varphi_2(a)\,|a|^{s-\frac{1}{2}}\, d^\times a = L(s,\mu_2)L(s,\nu_2) \;,$$
$$\int \pi_2(w)\varphi_2(a)\,|a|^{s-\frac{1}{2}}\omega_2^{-1}(a)\,d^\times a = L(s,\mu_2^{-1})L(s,\nu_2^{-1}) \;.$$

Now in (15.10.1) we see that the integrals in t and x are in fact extended to $t \in R^\times$ and $x \in R$. Hence

$$\omega(-1)\Psi(s,W_1,W_2,\Phi) =$$
$$\int_{F^\times \times R} \pi_1\left[w\begin{pmatrix}1 & x\\ 0 & 1\end{pmatrix}\right]\varphi_1(a)\,\pi_2\left[w\begin{pmatrix}1 & x\\ 0 & 1\end{pmatrix}\right]\varphi_2(-a)\,|a|^{s-1}\, d^\times a\,dx \;.$$

Now φ_2 is invariant under K. As for φ_1 its support is contained in $R - \{0\}$ and it is therefore invariant under

$$\pi_1\begin{pmatrix}1 & x\\ 0 & 1\end{pmatrix} \quad \text{where} \quad x \in R \;.$$

So we find

$$\omega(-1)\Psi(s,W_1,W_2,\Phi) = \int \pi_1(w)\varphi_1(a)\varphi_2(-a)\,|a|^{s-1}\, d^\times a \quad .$$

This integral is computed as a power series in q^{-s}. Using Lemma 15.10.5 we find its value to be

$$\omega(-1)\Psi(s,W_1,W_2,\Phi) = L(s,\mu_2\nu_1)L(s,\nu_2\nu_1)\epsilon(1,\mu_1^{-1})\nu_1(\varpi^u) \ .$$

Similarly in (15.10.2) we see that the integrals in t and x are extended, in fact, to $t \in \varpi^{-u}R^\times$ and $x \in \varpi^u R$. Hence

$$\omega(-1)\tilde{\Psi}(1-s,W_1,W_2,\hat{\Phi}) = \epsilon(2s-1,\omega) \times$$

$$\int_{F^\times \times R\varpi^u} \pi_1\left[w\begin{pmatrix}1 & x\\ 0 & 1\end{pmatrix}w\right]\varphi_1(a)\pi_2\left[w\begin{pmatrix}1 & x\\ 0 & 1\end{pmatrix}w\right]\varphi_2(-a)\,|a|^{-s}\,\omega^{-1}(a)d^\times a \ .$$

Again we use the fact that φ_2 is K invariant. On the other hand, we observe that the support of $\pi_1(w)\varphi_1$ is contained in the set $\varpi^{-u}R - \{0\}$. Hence it is invariant under

$$\pi_1\begin{pmatrix}1 & x\\ 0 & 1\end{pmatrix} \text{ where } x \in \varpi^u R \quad .$$

So we find this time

$$\omega(-1)\tilde{\Psi}(1-s,W_1,W_2,\hat{\Phi})$$

$$= \epsilon(2s-1,\omega)q^{-u}\omega_1(-1)\int\varphi_1(a)\varphi_2(a)\,|a|^{-s}\omega^{-1}(a)d^\times a$$

$$= \epsilon(2s-1,\omega)q^{-u}\omega_1(-1)L(1-s,\nu_1^{-1}\mu_2^{-1})L(1-s,\nu_1^{-1}\nu_2^{-1}) \quad .$$

The functional equation (15.2) is therefore equivalent to

$$\epsilon(2s-1,\omega)q^{-u}\omega(-1)/\epsilon(1,\mu_1^{-1})\nu_1(\varpi^u) = \epsilon(s,\mu_1\mu_2)\epsilon(s,\mu_1\nu_2) \ .$$

This is easily proved with the help of (15.10.3).

<u>Case 15.12</u>: We assume that $\pi_1 = \pi(\mu_1,\nu_1)$ and $\pi_2 = \pi(\mu_2,\nu_2)$ with

$$u_1 = O(\omega) = O(\mu_1\mu_2) = O(\mu_1) \geq u_2 = O(\mu_2) ,$$

$$O(\nu_1) = O(\nu_2) = 0 .$$

We choose again

$$\Phi(x,y) = \Phi_\omega(x)\Phi_0(y)$$

and W_i corresponding to φ_i such that, for $i = 1,2$,

$$\varphi_i(a\epsilon) = \varphi_i(a)\omega_i(\epsilon) , \text{ for } \epsilon \in R^\times ,$$

$$\int \varphi_i(a)\omega_i^{-1}(a) |a|^{s-\frac{1}{2}} d^\times a = L(s,\nu_i^{-1}) ,$$

$$\int \pi_i(w)\varphi_i(a) |a|^{s-\frac{1}{2}} d^\times a = L(s,\nu_i)\epsilon(1-s,\mu_i^{-1}) ,$$

$$= L(s,\nu_i)\epsilon(1,\mu_i^{-1})q^{u_i s} .$$

Then in (15.10.1) we integrate for t in $R^\times$ and x in R . We find in that way

$$\omega(-1)\Psi(s,W_1,W_2,\Phi)$$

$$= \int \pi_1(w)\varphi_1(a)\pi_2(w)\varphi_2(-a) |a|^{s-1} d^\times a$$

$$= \epsilon(1,\mu_1^{-1})\epsilon(1,\mu_2^{-1})L(s,\nu_1\nu_2)q^{u_2 s} \nu_1(\varpi^{u_1-u_2}) ,$$

the last line by application of Lemma 15.10.4.

Similarly, in (15.10.2) we integrate for t in $\varpi^{-u_1}R^\times$ and x in $\varpi^{u_1}R$. We find in that way

$$\omega(-1)\tilde{\Psi}(1-s,W_1,W_2,\hat{\Phi})$$

$$= q^{-u_1}\epsilon(2s-1,\omega)\omega(-1)\int\varphi_1(a)\varphi_2(-a) |a|^{-s} \omega^{-1}(a)d^\times a$$

$$= q^{-u_1}\epsilon(2s-1,\omega)\omega_1(-1)L(1-s,\nu_1^{-1}\nu_2^{-1}) .$$

The functional equation (15.2) is now equivalent to

$$q^{-u_1}\epsilon(2s-1,\omega)\omega(-1)/\epsilon(1,\mu_1^{-1})\epsilon(1,\mu_2^{-1})q^{u_2 s}\,\nu_1(\varpi^{u_1-u_2})$$

$$= \epsilon(s,\mu_1\mu_2)\epsilon(s,\mu_1\nu_2)\epsilon(s,\nu_1\mu_2)\,\epsilon(s,\nu_1\nu_2)\ .$$

which is easily checked by using (15.10.3).

Case 15.13: We assume that $\pi_1 = \pi(\mu_1,\nu_1)$ and $\pi_2 = \pi(\mu_2,\nu_2)$ with

$$u = 0(\mu_1) = 0(\mu_2) \geq v = 0(\omega) = 0(\mu_1\mu_2) > 0\ ,$$

$$0(\nu_1) = 0(\nu_2) = 0\ .$$

Again we take Φ such that

$$\Phi(x,y) = \Phi_\omega(x)\Phi_0(y)$$

and W_1 corresponds to φ_1 defined by the same conditions as in the previous case. We take W_2 corresponding to φ_2 defined by

$$\varphi_2(a\epsilon) = \varphi_2(a)\omega_2(\epsilon)\ ,\ \text{for}\quad \epsilon \in R^\times\ ,$$

$$\int \pi_2(w)\varphi_2(a)\,|a|^{s-\frac{1}{2}}d^\times a = L(s,\nu_2)\ ,$$

$$\int \varphi_2(a)\,|a|^{s-\frac{1}{2}}\omega_2^{-1}(a)d^\times a = L(s,\nu_2^{-1})\omega_2(-1)\epsilon(1-s,\mu_2)$$

$$= L(s,\nu_2^{-1})\epsilon(1,\mu_2)q^{us}\omega_2(-1)\ .$$

We find now

$$\omega(-1)\Psi(s,W_1,W_2,\Phi) = \int \pi_1(w)\varphi_1(a)\pi_2(w)f_2(-a)\,|a|^{s-1}\,d^\times a$$

where we set

$$f_2(a) = \int_R \pi_2\begin{pmatrix}1 & x\\ 0 & 1\end{pmatrix}\varphi_2(a)dx\ .$$

It is clear that f_2 belongs to $\mathcal{K}(\pi_2,\psi)$. Moreover it is given by

$$f_2(a) = \int_R \psi(ax)\varphi_2(a)dx = \varphi_2(a) \quad \text{if}\quad a \in R$$

$$= 0 \quad \text{otherwise.}$$

Hence

$$\int f_2(a)\,|a|^{s-\frac{1}{2}}\,\omega_2^{-1}(a)d^{\times}a = \epsilon(1,\mu_2)\omega_2(-1)L(s,\nu_2^{-1})\nu_2^{-1}(\varpi^u) .$$

The functional equation of the functions in the Kirillov model gives

$$\int \pi_2(w)f_2(a)\,|a|^{s-\frac{1}{2}}\,d^{\times}a = L(s,\nu_2)\nu_2^{-1}(\varpi^u)q^{u(s-1)} .$$

So we find

$$\omega(-1)\Psi(s,W_1,W_2,\Phi) = L(s,\nu_1\nu_2)q^{u(s-1)}\epsilon(1,\mu_1^{-1})\nu_2^{-1}(\varpi^u) .$$

In the same manner we get

$$\omega(-1)\widetilde{\Psi}(1-s,W_1,W_2,\hat{\Phi}) = \epsilon(2s-1,\omega)\int\pi_1(w)f_1(a)\varphi_2(a)\,|a|^{-s}\omega^{-1}(a)d^{\times}a .$$

Where we set

$$f_1(a) = \int_{\varpi^v R} \pi_1\begin{pmatrix}1 & x\\ 0 & 1\end{pmatrix}\pi_1(w)\varphi_1(a)dx .$$

Here f_1 is given explicitly by

$$f_1(a) = q^{-v}\pi_1(w)\varphi_1(a) \quad \text{if} \quad a \in \varpi^{-v}R$$

$$= 0 \quad \text{otherwise.}$$

Hence we find in succession

$$\int f_1(a)\,|a|^{s-\frac{1}{2}}d^{\times}a = L(s,\nu_1)q^{(s-1)v}\,\nu_1(\varpi^{u-v})\epsilon(1,\mu_1^{-1}) ,$$

$$\int\pi_1(w)f_1(a)\,|a|^{s-\frac{1}{2}}\omega_1^{-1}(a)d^{\times}a = L(s,\nu_1^{-1})q^{s(u-v)-u}\,\omega_1(-1)\nu_1(\varpi^{u-v}) ,$$

$$\omega(-1)\widetilde{\Psi}(1-s,W_1,W_2,\hat{\Phi})$$

$$= \omega(-1)\epsilon(2s-1,\omega)L(1-s,\nu_1^{-1}\nu_2^{-1})q^{(1-s)(u-v)-u}\nu_1(\varpi^{u-v})\nu_2(\varpi^{-v})\epsilon(1,\mu_2) .$$

Again it is a simple matter to check that the functional equation (15.2) is satisfied.

<u>Case 15.14</u>: We assume that $\pi_1 = \pi(\mu_1,\nu_1)$ and $\pi_2 = \pi(\mu_2,\nu_2)$ with

$$u = O(\mu_1) = O(\mu_2) > 0 = O(\mu_1\mu_2) = O(\omega) , \; O(\nu_1) = O(\nu_2) = 0 .$$

Now we take Φ such that

$$\Phi(x,y) = \Phi_{\mu_1}(x)\Phi_{\mu_2}(y) .$$

We take W_i corresponding to φ_i defined by the following conditions:

$$\varphi_1(a\varepsilon) = \varphi_1(a)\omega_1(\varepsilon) \quad \text{for} \quad \varepsilon \quad \text{in} \quad R^\times ,$$

$$\int \varphi_1(a)|a|^{s-\frac{1}{2}}\omega_1^{-1}(a)d^\times a = L(s,\nu_1^{-1}) ,$$

$$\int \pi_1(w)\varphi_1(a)|a|^{s-\frac{1}{2}}d^\times a = L(s,\nu_1)\varepsilon(1-s,\mu_1^{-1}) ;$$

$$\varphi_2(a\varepsilon) = \varphi_2(a) \quad \text{for} \quad \varepsilon \quad \text{in} \quad R^\times ,$$

$$\int \varphi_2(a)|a|^{s-\frac{1}{2}}d^\times a = L(s,\nu_2)\varepsilon(1-s,\mu_2^{-1}) ,$$

$$\int \pi_2(w)\varphi_2(a)|a|^{s-\frac{1}{2}}\omega_2^{-1}(a)d^\times a = L(s,\nu_2^{-1})\omega_2(-1) .$$

In $\Psi(s,W_1,W_2,\Phi)$ we integrate for t in $R^\times$ and x in $R^\times$. We find in that way

$$\omega(-1)\Psi(s,W_1,W_2,\Phi) = \int \pi_1(w)\varphi_1(a)\pi_2(w)f_2(-a)d^\times a$$

where f_2 is the element of $\mathcal{K}(\pi_2,\psi)$ defined by

$$f_2(a) = \int_{R^\times} \pi_2\begin{pmatrix}1 & x\\ 0 & 1\end{pmatrix}\mu_2^{-1}(x)dx \ \varphi_2(a) .$$

With the notations of [1] this can be written as

$$f_2(a) = \varphi_2(a)\eta(\mu_2^{-1},a)(1-q^{-1}) .$$

We remind the reader that if $d^\times x$ is the Haar measure of $F^\times$ for which $R^\times$ has measure one and μ a quasi-character of $F^\times$ we set

$$\eta(\mu,a) = \int_{R^\times} \mu(x)\psi(ax)d^\times x .$$

Then if the order of μ is $u > 0$ we have

$$\varepsilon(s,\mu) = (1-q^{-1})q^{u(1-s)}\mu(\varpi^u)\eta(\mu^{-1},\varpi^{-u}) .$$

It is easily seen that

$$\int f_2(a)\omega_2^{-1}(a)\,|a|^{s-\frac{1}{2}}d^\times a = q^{u(s-1)}\nu_2(\varpi^u)\mu_2(-1) .$$

Therefore

$$\int \pi_2(w)f_2(a)\,|a|^{s-\frac{1}{2}}d^\times a = \nu_2(\varpi^u)\mu_2(-1)\,\epsilon(1,\mu_2^{-1})L(s,\nu_2)L(1-s,\nu_2^{-1})^{-1} ,$$

and, by application of (15.10.4),

$$\omega(-1)\Psi(s,W_1,W_2,\Phi) =$$

$$\epsilon(1,\mu_1^{-1})\,\epsilon(1,\mu_2^{-1})\nu_1\nu_2(\varpi^u)\omega_2(-1)L(s,\nu_1\nu_2)L(1-s,\nu_1^{-1}\nu_2^{-1})^{-1} .$$

In $\widetilde{\Psi}$ we integrate for t in $R^\times\varpi^{-u}$ and x in $R^\times$. We first find, for x in $R^\times$,

$$\int \hat{\Phi}_{\mu_1}(t)\hat{\Phi}_{\mu_2}(tx)\,|t|^{2-2s}\omega^{-1}(t)d^\times t = \mu_2(x)\,\epsilon(s,\mu_1\nu_2)\,\epsilon(s,\mu_2\nu_1) .$$

Then we get

$$\omega(-1)\widetilde{\Psi}(1-s,W_1,W_2,\hat{\Phi}) =$$

$$\epsilon(s,\mu_1\nu_2)\,\epsilon(s,\mu_2\nu_1)\omega_2(-1)\int \pi_1(w)f_1(a)\varphi_2(a)\omega^{-1}(a)\,|a|^{-s}d^\times a$$

where we set

$$f_1(a) = \int_{R^\times} \pi_1\begin{pmatrix}1 & x\\ 0 & 1\end{pmatrix}\pi_1(w)\varphi_1(a)\mu_2(x)dx .$$

This time we obtain

$$\int f_1(a)\,|a|^{s-\frac{1}{2}}\omega_2(a)d^\times a = q^{us}\,\epsilon(1,\mu_1^{-1})\,\epsilon(1,\mu_2^{-1})\nu_2(\varpi^{-u})$$

then

$$\int \pi_1(w)f_1(a)\,|a|^{s-\frac{1}{2}}\omega^{-1}(a)d^\times a =$$

$$\epsilon(0,\nu_1\mu_2)\,\epsilon(1,\mu_1^{-1})\,\epsilon(1,\mu_2^{-1})L(s,\mu_1^{-1}\mu_2^{-1}\nu_2^{-1})L(1-s,\mu_1\mu_2\nu_2)^{-1} ,$$

and at last

$$\omega(-1)\widetilde{\Psi}(1-s,W_1,W_2,\hat{\Phi}) =$$

$$\nu_1\nu_2(\varpi^u)\,\epsilon(1,\mu_1^{-1})\,\epsilon(1,\mu_2^{-1})\,\epsilon(s,\mu_1\nu_2)\,\epsilon(s,\mu_2\nu_1)L(s,\mu_1\mu_2)L(1-s,\mu_1^{-1}\mu_2^{-1})^{-1} .$$

The functional equation (15.2) is clearly satisfied.

We now take up case (15.15).

Case 15.15: We assume that π_1 is absolutely cuspidal or $\pi_1 = \pi(\mu_1,\nu_1)$, and all the products

$$\mu_1\nu_2\ ,\ \mu_1\mu_2\ ,\ \nu_1\mu_2\ ,\ \nu_1\nu_2\ ,$$

are ramified.

For φ_i in $\mathcal{K}(\pi_i,\psi)$, n in $\underline{Z}$ and all quasi-characters χ of $F^\times$ (or characters of $R^\times$) we set

$$\hat{\varphi}_i(n,\chi) = \int_{R^\times} \varphi_i(\epsilon\varpi^n)\chi(\epsilon)d\epsilon$$

where $d\epsilon$ is the normalized Haar measure of $R^\times$. The formal Mellin transform of φ_i is the family of formal series

$$\hat{\varphi}_i(\chi,X) = \sum \hat{\varphi}_i(n,\chi)X^n\ .$$

The operator $\pi_i(w)$ is defined by the formula

$$\widehat{\pi_i(w)\varphi_i}(\chi,X) = C_i(\chi,X)\hat{\varphi}_i(\chi^{-1}\omega_i^{-1},X^{-1}\omega_i^{-1}(\varpi))\ ,$$

where

$$C_i(\chi,X) = \sum C_i(n,\chi)X^n$$

is a suitable family of formal series. (Note that for convenience the notations of [1] have been slightly changed).

There are two integers $c \geq 2$ and $d \geq 2$ so that $\epsilon(s,\pi_1 \otimes \mu_2) = q^{-cs}\epsilon(0,\pi_1 \otimes \mu_2)$, $\epsilon(s,\pi_1 \otimes \nu_2) = q^{-ds}\epsilon(0,\pi_1 \otimes \nu_2)$.

The definition of c and d amounts to say that

$$C_1(n,\mu_2) \neq 0 \iff n = -c\ ,$$

$$C_1(n,\nu_2) \neq 0 \iff n = -d\ .$$

We set

$$z = O(\omega) \ , \ v = O(\mu_2 . \nu_2^{-1}) \ .$$

We claim that we need to treat only the following subcases.

(15.16) $\quad z = 0 \ , \ c = d \geq v + 1 \ ,$

(15.17) $\quad z > 0 \ , \ c \geq v + 1 \ , \ d \geq v + 1 \ , \ c \geq z + 1 \ , \ d \geq z + 1 \ ,$

(15.18) $\quad z = v > 0 \ , \ c \leq v \ , \ d = 2z \ .$

Indeed suppose first that $z = 0$; then ω is unramified. The relation $\pi_1(w)^2 = \omega_1(-1)$ gives, for all characters χ of $R^\times$,

$$C_1(\chi,X)C_1(\chi^{-1}\omega_1^{-1},X^{-1}\omega_1^{-1}(\varpi)) = \omega_1(-1) \ .$$

For $\chi = \mu_2$, this implies that $c = d$ and

$$C_1(\mu_2,-c)C_1(\nu_2,-d) = \omega_1(-1)\omega_1(\varpi^{-c}) \ .$$

The relation 2.11.ii of [1] reduces then, for $n = p = 1 - c$, to

$$\sum \eta(\sigma^{-1}\mu_2,\varpi^{-c+1})\eta(\sigma^{-1}\nu_2,\varpi^{-d+1})C_1(2-2c,\sigma) = \omega_1(-\varpi^{1-c}) \ / \ (1-q) \ .$$

Now we recall that $\eta(\mu,\varpi^n)$ vanishes unless

$$n = -O(\mu) \quad \text{if} \quad O(\mu) > 0 \ ,$$

$$n \geq -1 \quad \text{if} \quad O(\mu) = 0 \ .$$

If $c > 2$ we conclude that there is at least one character σ of $R^\times$ such that

$$O(\sigma^{-1}\mu_2) = O(\sigma^{-1}\nu_2) = c - 1 \ .$$

If $c = 2$ we conclude that there is at least one character σ of $R^\times$ such that

$$O(\sigma^{-1}\mu_2) = 0 \ \text{or} \ 1 \ , \ O(\sigma^{-1}\nu_2) = 0 \ \text{or} \ 1 \ .$$

In any case the order of

$$\mu_2 . \nu_2^{-1} = \sigma^{-1}\mu_2(\sigma^{-1}\nu_2)^{-1}$$

is at most $c - 1$. Hence we are in case (15.16).

Assume now that $z > 0$. This time the relation (2.11.i) of [1] gives for $n = c + z$ and $p = -d + z$

$$\sum \eta(\sigma^{-1}\mu_2, \varpi^{-c+z})\eta(\sigma^{-1}\nu_2, \varpi^{-d+z})C_1(-c-d+2z, \sigma)$$
$$= \omega_1(\varpi^{z-c})\omega_2(-1)\eta(\omega^{-1}, \varpi^{-z}) \quad .$$

If $c - z \leq 0$ then the only nonzero term on the left-hand side is for $\sigma = \mu_2$. So we find that

$$\eta(\nu_2\mu_2^{-1}, \varpi^{-d+z})C_1(-c-d+2z, \mu_2) \neq 0 \ .$$

This implies that

$$-c-d+2z = -c$$

that is $d = 2z$. Since the order of $\mu_2\nu_2^{-1}$ is v it implies also that either $v = z$ or $v = 0$ and $z = 1$. The relations

$$z = 1 \ , \ c \geq 2 \ \text{ and } \ c - z \leq 0$$

are not compatible. Therefore $z = v$ and we are in case (15.18).

Similarly, if $d - z \leq 0$ we are in case (15.18) with the roles of μ_2 and ν_2 exchanged.

So we may assume $c > z$, $d > z$, and $z > 0$. If $c \leq v$ then we get from (2.11.i) in [1]

$$\sum \eta(\sigma^{-1}\mu_2, \varpi^{-c+v})\eta(\sigma^{-1}\nu_2^{-1}\omega_1^{-1}, \varpi^{-p+v})C_1(-c-p+2v, \sigma)$$
$$= C_1(-c, \mu_2)C_1(-p, \nu_2^{-1}\omega_1^{-1})\eta(\nu_2\mu_2^{-1}, \varpi^{-v})\omega_1(\varpi^v)\mu_2\nu_2^{-1}(-1) \ .$$

We choose p so that $C_1(-p, \nu_2^{-1}\omega_1^{-1}) \neq 0$. Then the only nonzero term on the left-hand side is for $\sigma = \mu_2$. So we get

$$\eta(\omega^{-1}, \varpi^{-p+v})C_1(-c-p+2v, \mu_2) \neq 0 \ .$$

This implies that

$$p = 2v \quad \text{and} \quad v - p = -z \quad .$$

Hence $z = v \geq c$, which is a contradiction. Therefore we have proved that $c > v$. Similarly $d > v$. So we are in case (15.17).

Finally there is no harm in replacing the pair (π_1, π_2) by the pair $(\pi_1 \otimes \chi, \pi_2 \otimes \chi^{-1})$ where χ is a quasi-character of $F^\times$. In particular, by taking the order of χ to be large enough, we may assume furthermore that:

$$L(s,\pi_1) = L(s,\tilde{\pi}_1) = L(s,\pi_2) = L(s,\tilde{\pi}_2) = 1$$

and

$$\epsilon(s,\pi_i) = q^{-2su}\epsilon(0,\pi_i) \ , \ O(\mu_2) = O(\nu_2) = u \ ,$$

where u is an integer strictly larger than c,d,z and v . (Cf. [1], Proposition 3.8).

The key to the functional equation (15.2) will be the following lemma.

<u>Lemma 15.15.1</u>: <u>With the above assumptions let</u> φ_i <u>be the element of</u> $\mathcal{K}(\pi_i,\psi)$ <u>defined by the following conditions</u>:

$$\varphi_i(a\epsilon) = \varphi_i(a)\omega_i(\epsilon) \quad \text{for} \quad \epsilon \in R^\times \ ,$$

$$\int \varphi_i(a)\,|a|^{s-\frac{1}{2}}\omega_i^{-1}(a)d^\times a = \epsilon(1-s,\pi_i) \quad .$$

<u>Let</u> $d^\times a$ <u>be the Haar measure of</u> $F^\times$ <u>for which the measure of</u> $R^\times$ <u>is one and</u> dx <u>be the self dual Haar measure on</u> F . <u>Then the integral</u>

$$J = \int_{F^\times \times R} \pi_1\left[w\begin{pmatrix}1 & x\\ 0 & 1\end{pmatrix}\right]\varphi_1(a)\pi_2\left[w\begin{pmatrix}1 & x\\ 0 & 1\end{pmatrix}\right]\varphi_2(-a)\,|a|^{s-1}d^\times a\,dx$$

<u>is convergent for</u> $\mathrm{Re}\,s$ <u>large enough</u>. <u>In case</u> (15.16) <u>and</u> (15.17) <u>its value is given by</u>

$$\varepsilon(1,\pi_1)\varepsilon(1,\pi_2)(1-q^{-1})$$

$$\times \sum_{1\le a,b\le 2u} \mu_2(\varpi^{-a})\nu_2(\varpi^{-b})q^{(a+b-2u)(s-1)+\frac{1}{2}(a+b)}$$

$$\times \sum_{\chi} C_1(-a-b,\chi^{-1})\eta(\chi\mu_2,\varpi^{-a})\eta(\chi\nu_2,\varpi^{-b}) \ .$$

In case (15.18), its value is

$$\varepsilon(1,\pi_1)\varepsilon(1,\pi_2)(1-q^{-1})q^{(c-2u)(s-1)+\frac{1}{2}c}\ \mu_2(\varpi^{v-c})\nu_2(\varpi^{-v})$$

$$\times \sum_{\chi} C_1(-c,\chi)\eta(\mu_2\chi^{-1},\varpi^{-c+v})\eta(\nu_2\chi^{-1},\varpi^{-v}) \ .$$

In both expressions we sum on all characters χ of the compact group $R^{\times}$.

The fact that the integral converges for $\operatorname{Re} s$ large enough is readily seen. Using Fourier theorem we see that the value of J is

$$\int_R dx \sum_{\chi,n} \chi(-1)q^{-n(s-1)}\widehat{\pi_1\left[w\begin{pmatrix}1 & x\\ 0 & 1\end{pmatrix}\right]\varphi_1(n,\chi)\pi_2\left[w\begin{pmatrix}1 & x\\ 0 & 1\end{pmatrix}\right]\varphi_2}(n,\chi^{-1}) \ .$$

Now

$$\widehat{\pi_i\begin{pmatrix}1 & x\\ 0 & 1\end{pmatrix}\varphi_i}(n,\chi) = \varphi_i(\varpi^n)\eta(\chi\omega_i,\varpi^n x)$$

and

$$\widehat{\pi_i(w)\pi_i\begin{pmatrix}1 & x\\ 0 & 1\end{pmatrix}\varphi_i}(n,\chi) = \sum C_i(m+n,\chi)\varphi_i(\varpi^m)\eta(\chi^{-1},\varpi^m x)\omega_i(\varpi^{-m}) \ .$$

But here φ_i is so chosen that

$$\varphi_i(\varpi^n) = 0 \quad \text{unless} \quad n = -2u \ ,$$

$$\varphi_i(\varpi^{-2u})q^{-u}\omega_i(\varpi^{2u}) = \varepsilon(1,\pi_i) \ .$$

Therefore

$$\widehat{\pi_i\left[w\begin{pmatrix}1 & x\\ 0 & 1\end{pmatrix}\right]\varphi_i}(n,\chi) = \varepsilon(1,\pi_i)q^u C_i(n-2u,\chi)\eta(\chi^{-1},\varpi^{-2u}x) \ .$$

Accordingly, we find

$$J\epsilon(1,\pi_1)^{-1}\epsilon(1,\pi_2)^{-1} =$$

$$\sum_{\chi,n} C_1(n-2u,\chi)C_2(n-2u,\chi^{-1})\chi(-1)q^{2u}\int_R \eta(\chi^{-1},x\varpi^{-2u})\eta(\chi,x\varpi^{-2u})dx .$$

But, by a change of variable, we find

$$\chi(-1)q^{2u}\int_R \eta(\chi^{-1},x\varpi^{-2u})\eta(\chi,x\varpi^{-2u})dx = \chi(-1)\int_{\varpi^{-2u}R}\eta(\chi^{-1},x)\eta(\chi,x)dx .$$

This integral vanishes unless the order of χ satisfies

$$0 \le O(\chi) \le 2u .$$

If we assume that $O(\chi)$ satisfies this inequality, the integral is not changed if we replace the domain of integration by the whole group F. Then it can be evaluated by using Fourier theorem. Its value is then found to be $(1-q^{-1})^{-1}$. Reporting this in the above expression for J we obtain:

(15.15.2) $$J\epsilon(1,\pi_1)^{-1}\epsilon(1,\pi_2)^{-1}(1-q^{-1}) =$$

$$\sum_{n,0\le O(\chi)\le 2u} C_1(n-2u,\chi^{-1})C_2(n-2u,\chi)q^{-n(s-1)} .$$

The sum is extended to all n in $\underline{Z}$ and all characters χ of $R^\times$ whose order is less than or equal to $2u$.

Assume first that $\mu_2\nu_2^{-1}$ is unramified, i.e., that $v = 0$. Then we are in case (15.16) or (15.17). Since we assume that the order of μ_2 is u we see that the inequalities

$$O(\chi\mu_2) \le 2u \quad \text{and} \quad O(\chi) \le 2u$$

are equivalent. (By $\chi\mu_2$ we denote the product of χ by the restriction of μ_2 to $R^\times$). In (15.15.2) we single out the term corresponding to $\chi = \mu_2^{-1} = \nu_2^{-1}$. We obtain in that way

$$J\epsilon(1,\pi_1)^{-1}\epsilon(1,\pi_2)^{-1}(1-q^{-1}) = \sum_n C_1(n-2u,\mu_2)C_2(n-2u,\mu_2^{-1})q^{-n(s-1)}$$

$$+ \sum_{n,1\leq a\leq 2u,} \ \sum_{O(\chi\mu_2)=a} C_1(n-2u,\chi^{-1})C_2(n-2u,\chi)q^{-n(s-1)} \quad .$$

But $C_1(n,\mu_2)$ is different of zero if and only if $n = -c$. Moreover, $\chi\mu_2$ and $\chi\nu_2$ are equal. If

$$O(\mu_2\chi) = O(\nu_2\chi) = a \geq 1$$

then $C_2(n,\chi)$ is nonzero if and only if $n = -2a$ and

$$C_2(-2a,\chi) = (1-q^{-1})^2 q^a \omega_2(\varpi^{-a})\eta(\chi\mu_2,\varpi^{-a})\eta(\chi\nu_2,\varpi^{-a}) \ .$$

Taking those facts into account we find

$$J\epsilon(1,\pi_1)^{-1}\epsilon(1,\pi_2)^{-1}(1-q^{-1}) = C_1(-c,\mu_2)C_2(-c,\mu_2^{-1})q^{(c-2u)(s-1)}$$

$$+ \sum_{1\leq a\leq 2u} (1-q^{-1})^2\omega_2(\varpi^{-a})q^{(2a-2u)(s-1)+a}$$

$$\times \sum_{O(\chi\mu_2)=a} C_1(-2a,\chi^{-1})\eta(\chi\mu_2,\varpi^{-a})\eta(\chi\nu_2,\varpi^{-a}) \quad .$$

Note that

$$\eta(\chi\mu_2,\varpi^{-a})(\chi\nu_2,\varpi^{-a}) \neq 0$$

implies

$$O(\chi\mu_2) = O(\chi\nu_2) = a \quad \text{if} \quad a > 1$$

$$O(\chi\mu_2) = O(\chi\nu_2) = 1 \quad \text{or} \quad 0 \quad \text{if} \quad a = 1 \ .$$

So, in the above expression, we may sum for all characters χ of $R^{\times}$, provided we subtract the term corresponding to

$$a = 1 \ , \ \chi\mu_2 = \chi\nu_2 = 1 \quad .$$

We obtain in that way

$$J(1-q^{-1})\varepsilon(1,\pi_1)^{-1}\varepsilon(1,\pi_2)^{-1} = C_1(-c,\mu_2)C_2(-c,\mu_2^{-1})q^{(c-2u)(s-1)}$$

$$- C_1(-2,\mu_2)(1-q^{-1})^2\omega_2(\varpi^{-1})\eta(1,\varpi^{-1})^2\ q^{(2-2u)(s-1)+1}$$

$$+ (1-q^{-1})^2 \sum_{1\le a\le 2u} \omega_2(\varpi^{-a})q^{(2a-2u)(s-1)+a}$$

$$\times \sum_{\chi} C_1(-2a,\chi^{-1})\eta(\chi\mu_2,\varpi^{-a})\eta(\chi\nu_2,\varpi^{-a}) \ .$$

Now we remind ourselves that

$$C_1(n,\mu_2) \neq 0 \iff n = -c \ ,$$

$$C_2(n,\mu_2^{-1}) \neq 0 \iff n \geq -2 \ .$$

In any case $c \geq 2$. If $c > 2$ then

$$C_1(-2,\mu_2) = 0 \quad \text{and} \quad C_2(-c,\mu_2^{-1}) = 0 \ .$$

So the two first terms disappear. If $c = 2$ then taking into account that

$$\eta(1,\varpi^{-1})^2 = q^{-2}(1-q^{-1})^{-2} \ ,$$

$$C_2(-2,\mu_2^{-1}) = q^{-1}\omega_2(\varpi^{-1}) \ ,$$

we see that the two first terms mutually cancel. So we are left with

$$J\varepsilon(1,\pi_1)^{-1}\varepsilon(1,\pi_2)^{-1}(1-q^{-1}) = \sum_{1\le a\le 2u} \omega_2(\varpi^{-a})q^{(2a-2u)(s-1)+a}$$

$$\times \sum_{\chi} C_1(-2a,\chi^{-1})\eta(\chi\mu_2,\varpi^{-a})\eta(\chi\nu_2,\varpi^{-a}) \ .$$

Since we have assumed that μ_2 and ν_2 coincide on $R^{\times}$, we see that for $a \geq 1$ and $b \geq 1$ the relation

$$\eta(\chi\mu_2,\varpi^{-a})\eta(\chi\nu_2,\varpi^{-b}) \neq 0$$

implies $a = b$. Therefore the above expression can also be written as

$$J\epsilon(1,\pi_1)^{-1}\epsilon(1,\pi_2)^{-1}(1-q^{-1}) = \sum_{1\le a,b\le 2u} \mu_2(\varpi^{-a})\nu_2(\varpi^{-b})q^{(a+b-2u)+\frac{1}{2}(a+b)}$$

$$\times \sum_{\chi} C_1(-a-b,\chi^{-1})\eta(\chi\mu_2,\varpi^{-a})\eta(\chi\nu_2,\varpi^{-b}) .$$

So we have proved Lemma 15.15.1 when $\mu_2\nu_2^{-1}$ is unramified.

We assume now that $\mu_2\nu_2^{-1}$ is ramified, i.e., that $v > 0$. The method is the same as above but a little more complicated. We go back to (15.15.2). This time we single out the terms corresponding to

$$\chi = \mu_2^{-1} , \chi = \nu_2^{-1} .$$

We obtain in that way

$$J(1-q^{-1})\epsilon(1,\pi_1)^{-1}\epsilon(1,\pi_2)^{-1} = \sum_n C_1(n-2u,\mu_2)C_2(n-2u,\mu_2^{-1})q^{-n(s-1)}$$

$$+\sum_n C_1(n-2u,\nu_2)C_2(n-2u,\nu_2^{-1})q^{-n(s-1)}$$

$$+ \sum_{n,1\le a,b\le 2u} q^{-n(s-1)} \sum_{0(\chi\mu_2)=a,0(\chi\nu_2)=b} C_1(n-2u,\chi^{-1})C_2(n-2u,\chi) .$$

Once more we have to use the fact that

$$C_1(n,\mu_2) \neq 0 \iff n = -c ,$$

$$C_1(n,\nu_2) \neq 0 \iff n = -d ,$$

and that if

$$0(\chi\mu_2) = a \ge 1 , 0(\chi\nu_2) = b \ge 1 ,$$

then

$$C_2(n,\chi) \neq 0 \iff n = -a-b ,$$

$$C_2(-a-b,\chi) = (1-q^{-1})^2\mu_2(\varpi^{-a})\nu_2(\varpi^{-b})\eta(\chi\mu_2,\varpi^{-a})\eta(\chi\nu_2,\varpi^{-b}) .$$

So in the above expression we are left with

$$C_1(-c,\mu_2)C_2(-c,\mu_2^{-1})q^{(c-2u)(s-1)} + C_1(-d,\nu_2)C_2(-d,\nu_2^{-1})q^{(d-2u)(s-1)}$$

$$+ (1-q^{-1})^2 \sum_{1\le a,b\le 2u} \mu_2(\varpi^{-a})\nu_2(\varpi^{-b})q^{(a+b-2u)(s-1)+\frac{1}{2}(a+b)}$$

$$\times \sum_{O(\chi\mu_2)=a,\,O(\chi\nu_2)=b} C_1(-a-b,\chi^{-1})\eta(\chi\mu_2,\varpi^{-a})\eta(\chi\nu_2,\varpi^{-b}) \quad .$$

We may sum on all characters χ of $R^\times$ provided we subtract the terms corresponding to

$$a = 1 \ , \ b = v \ , \ \chi = \mu_2^{-1}$$

and

$$a = v \ , \ b = 1 \ , \ \chi = \nu_2^{-1} \ .$$

In that way we are able to express

$$J(1-q^{-1})\epsilon(1,\pi_1)^{-1}\epsilon(1,\pi_2)^{-1}$$

as the sum of

(15.15.3)

$$(1-q^{-1})^2 \sum_{1\le a,b\le 2u} \mu_2(\varpi^{-a})\nu_2(\varpi^{-b})q^{(a+b-2u)(s-1)+\frac{1}{2}(a+b)}$$

$$\times \sum_{\chi} C_1(-a-b,\chi^{-1})\eta(\chi\mu_2,\varpi^{-a})\eta(\chi\nu_2,\varpi^{-b})$$

and

(15.15.4)

$$C_1(-c,\mu_2)C_2(-c,\mu_2^{-1})q^{(c-2u)(s-1)} + C_1(-d,\nu_2)C_2(-d,\nu_2^{-1})q^{(d-2u)(s-1)}$$

$$-C_1(-1-v,\mu_2)\eta(1,\varpi^{-1})\eta(\mu_2\nu_2^{-1},\varpi^{-v})\mu_2(\varpi^{-1})\nu_2(\varpi^{-v})$$

$$\times(1-q^{-1})^2q^{(1+v-2u)(s-1)+\frac{1}{2}(v+1)}$$

$$-C_1(-1-v,\nu_2)\eta(1,\varpi^{-1})\eta(\nu_2\mu_2^{-1},\varpi^{-v})\mu_2(\varpi^{-v})\nu_2(\varpi^{-1})$$

$$\times(1-q^{-1})^2q^{(1+v-2u)(s-1)+\frac{1}{2}(v+1)} \quad .$$

Assume that we are in case (15.16) or (15.17). Then $c \geq v + 1$. On the other hand we know that

$$C_2(n,\mu_2^{-1}) \neq 0 \iff n \geq -v-1 ,$$

$$C_1(n,\mu_2) \neq 0 \iff n = -c .$$

If $c > v + 1$ we see that the first and third terms in (15.15.4) vanish. If on the contrary, $c = v + 1$, using the relations

$$\eta(1,\varpi^{-1}) = -q^{-1}(1-q^{-1})^{-1} ,$$

$$C_2(-1-v,\mu_2^{-1}) = (-1)(1-q^{-1})q^{\frac{1}{2}(v-1)}\mu_2(\varpi^{-1})\nu_2(\varpi^{-1})\eta(\mu_2\nu_2^{-1},\varpi^{-v}) ,$$

we see that the first and third terms mutually cancel. Similarly, we see that the second and fourth term in (15.15.4) vanish or cancel. Altogether the whole expression (15.15.4) vanish. Therefore

$$J(1-q^{-1})\varepsilon(1,\pi_1)^{-1}\varepsilon(1,\pi_2)^{-1}$$

reduces to the expression (15.15.3). Hence we have proved Lemma (15.15.1) in cases (15.16) and (15.17). It remains to examine the case (15.18). We assume therefore that

$$z = v \geq c \geq 2 , \quad d = 2z .$$

Again we write

$$J(1-q^{-1})\varepsilon(1,\pi_1)^{-1}\varepsilon(1,\pi_2)^{-1}$$

as the sum of (15.15.3) and (15.15.4). But in (15.15.3) we use the relation 2.11.i of [1] to see that

$$\sum_{\chi} C_1(-a-b,\chi^{-1})\eta(\chi\mu_2,\varpi^{-a})\eta(\chi\nu_2,\varpi^{-b})$$

$$= \eta(\omega,\varpi^{-z})\omega_1(\varpi^{z})\omega(-1)C_1(-a-z,\mu_2)C_1(-b-z,\nu_2) .$$

This is zero unless

$$a + z = c .$$

But this relation is incompatible with the assumption that

$$z = v \geq c ,$$

if, as in (15.15.3), the integer a is strictly positive. Therefore the whole expression (15.15.3) vanishes. In (15.15.4) we observe that

$$d = 2z \geq v + c > v + 1$$

implies

$$C_2(-d, \nu_2^{-1}) = 0 .$$

Similarly

$$c \leq v < v + 1$$

implies

$$C_1(-v-1, \mu_2) = 0$$

and

$$d = 2z > v + 1$$

implies

$$C_1(-v-1, \nu_2) = 0 .$$

We are therefore left with the expression

$$J(1-q^{-1})\epsilon(1,\pi_1)^{-1}\epsilon(1,\pi_2)^{-1} = C_1(-c,\mu_2)C_2(-c,\mu_2^{-1})q^{(c-2u)(s-1)} ,$$

or, substituting the value of $C_2(-c,\mu_2^{-1})$,

$$J = \epsilon(1,\pi_1)\epsilon(1,\pi_2)(1-q^{-1})q^{(c-2u)(s-1)+\frac{1}{2}c} \mu_2(\varpi^{v-c})\nu_2(\varpi^{-v})$$

$$\times C_1(-c,\mu_2)\eta(\mu_2\mu_2^{-1},\varpi^{v-c})\eta(\nu_2\mu_2^{-1},\varpi^{-v}) .$$

Since $v - c$ is positive or zero, we know that

$$\eta(\mu_2\chi,\varpi^{v-c})\eta(\nu_2\chi,\varpi^{-v}) \neq 0$$

if and only if $\chi = \mu_2^{-1}$. So we may rewrite the above formula in the form given in Lemma 15.15.1. This completes the proof of the lemma.

We check now the functional equation (15.2) for one choice of (W_1,W_2,Φ) .

<u>Cases</u> (15.17) <u>and</u> (15.18): In both cases the quasi-character ω is ramified with order $z > 0$. We choose Φ as follows:

$$\Phi(x,y) = \Phi_\omega(x)\Phi_0(y) .$$

We take W_i corresponding to φ_i in $\mathcal{K}(\pi_i,\psi)$ defined by the conditions of Lemma 15.15.1. To compute Ψ we use the formula (15.10.1). We integrate for t in $R^\times$ and x in R. We find then that

$$\omega(-1)\Psi(s,W_1,W_2,\Phi)$$

is just the integral J of Lemma 15.15.1.

If we are in case (15.17) we use the lemma to find

$$\omega(-1)\Psi(s,W_1,W_2,\Phi) = \epsilon(1,\pi_1)\epsilon(1,\pi_2)(1-q^{-1})$$

$$\times \sum_{1\le a,b\le 2u} \mu_2(\varpi^{-a})\nu_2(\varpi^{-b})q^{(a+b-2u)(s-1)+\frac{1}{2}(a+b)}$$

$$\times \sum_{\chi} C_1(-a-b,\chi^{-1})\eta(\chi\mu_2,\varpi^{-a})\eta(\chi\nu_2,\varpi^{-b}) .$$

By (2.11.i) of [1] the last sum is proportional to

$$C_1(-a-z,\mu_2)C_1(-b-z,\nu_2)$$

which is zero unless

$$a = c - z , \quad b = d - z .$$

Replacing the last sum by its expression and taking into account this remark we find

$$\omega(-1)\Psi(s,W_1,W_2,\Phi) = \epsilon(1,\pi_1)\epsilon(1,\pi_2)(1-q^{-1})q^{(c+d-2z-2u)(s-1)+\frac{1}{2}(c+d-2z)}$$

$$\omega(\varpi^{z})\mu_2(\varpi^{-c})\nu_2(\varpi^{-d})\omega(-1)\eta(\omega^{-1},\varpi^{-z})C_1(-c,\mu_2)C_1(-d,\nu_2).$$

Changing notations (cf. formula (2.18.1 of [1])) we find

(15.17.1) $$\Psi(s,W_1,W_2,\Phi)$$

$$= \epsilon(1,\pi_1)\epsilon(1,\pi_2)q^{2u(1-s)}\epsilon(2s,\omega)\epsilon(1-s,\tilde{\pi}_2\otimes\mu_2^{-1})\epsilon(1-s,\tilde{\pi}_2\otimes\nu_2^{-1}).$$

In case (15.18) we use Lemma 15.15.1 to find

$$\omega(-1)\Psi(s,W_1,W_2,\Phi)$$
$$= \epsilon(1,\pi_1)\epsilon(1,\pi_2)(1-q^{-1})q^{(c-2u)(s-1)+\frac{1}{2}c}\mu_2(\varpi^{v-c})\nu_2(\varpi^{-v})$$
$$\sum_{\chi} C_1(-c,\chi)\eta(\mu_2\chi^{-1},\varpi^{v-c})\eta(\nu_2\chi^{-1},\varpi^{-v})$$

or using (2.11.i) in [1]

$$= \epsilon(1,\pi_1)\epsilon(1,\pi_2)(1-q^{-1})q^{(c-2u)(s-1)+\frac{1}{2}c}\mu_2(\varpi^{v-c})\nu_2(\varpi^{-v})$$
$$\eta(\omega^{-1},\varpi^{-z})\omega_1(\varpi^{z})\omega(-1)C_1(-c,\mu_2)C_1(-d,\nu_2) \quad .$$

Changing notations we find again (15.17.1).

To compute $\tilde{\Psi}$ we use (15.10.2). We see that we have to integrate for t in $\varpi^{-z}R^{\times}$ and x in $\varpi^{z}R$. We obtain

$$\omega(-1)\tilde{\Psi}(1-s,W_1,W_2,\hat{\Phi}) =$$

$$\epsilon(2s-1,\omega)\int_{\varpi^{z}R}\pi_1\left[w\begin{pmatrix}1 & x\\ 0 & 1\end{pmatrix}w\right]\varphi_1(a)\pi_2\left[w\begin{pmatrix}1 & x\\ 0 & 1\end{pmatrix}w\right]\varphi_2(-a)\,|a|^{-s}\omega^{-1}(a)d^{\times}a\,dx.$$

But the support of $\pi_i(w)\varphi_i$ is $R^{\times}$. Therefore $\pi_i(w)\varphi_i$ is invariant under

$$\pi_i\begin{pmatrix}1 & x\\ 0 & 1\end{pmatrix}, \text{ for } x \text{ in } \varpi^{z}R .$$

We obtain therefore

$$\tilde{\Psi}(1-s,W_1,W_2,\hat{\Phi}) = q^{-z}\epsilon(2s-1,\omega)\int \varphi_1(a)\varphi_2(-a)\,|a|^{-s}\omega^{-1}(a)d^{\times}a$$

or

(15.17.2)
$$\tilde{\Psi}(1-s,W_1,W_2,\hat{\Phi}) = \omega_2(-1)\epsilon(2s,\omega)q^{2u(1-s)}\epsilon(1,\pi_1)\epsilon(1,\pi_2) \ .$$

Comparing (15.17.1) and (15.17.2) and taking into account the relations

$$\epsilon(s,\pi_2 \otimes \mu_2)\epsilon(1-s,\tilde{\pi}_2 \otimes \mu_2^{-1}) = \omega(-1)$$

$$\epsilon(s,\pi_2 \otimes \nu_2)\epsilon(1-s,\tilde{\pi}_2 \otimes \nu_2^{-1}) = \omega(-1)$$

we find

$$\Psi(s,W_1,W_2,\Phi)\epsilon(s,\pi_1 \otimes \mu_2)\epsilon(s,\pi_1 \otimes \nu_2) = \omega_2(-1)\tilde{\Psi}(1-s,W_1,W_2,\hat{\Phi}) ,$$

which is precisely the functional equation (15.2).

Case (15.16): In this case the quasi-character ω is unramified and has the form

$$\omega(x) = |x|^{\sigma} \quad \text{where} \quad \sigma \in \underline{C} .$$

We take

$$\Phi(x,y) = \Phi_0(x)\Phi_0(y) ,$$

and W_i corresponding to φ_i defined by the conditions of Lemma 15.15.1. To compute Ψ we use the formula

$$\Psi(s,W_1,W_2,\Phi) = \int_{F^{\times}\times K} \pi_1(k)\varphi_1(a)\pi_2(k)\varphi_2(-a)z(\alpha^{2s}\omega,k.\Phi)|a|^{s-1}d^{\times}a\,dk .$$

There is a similar formula for $\tilde{\Psi}$. As here $\Phi = \hat{\Phi}$ we see that

$$\tilde{\Psi}(1-s,W_1,W_2,\hat{\Phi}) = \Psi(1-s-\sigma,W_1,W_2,\Phi) .$$

So all we have to do is compute Ψ.

As Φ is K invariant, we have

$$z(\alpha^{2s}\omega,k.\Phi) = L(2s,\omega) .$$

Therefore

$$\Psi(s,W_1,W_2,\Phi) = L(2s,\omega)\int_{K\times F^{\times}} \pi_1(k)\varphi_1(a)\pi_2(k)\varphi_2(-a)|a|^{s-1}d^{\times}a\,dk .$$

After performing first the integration on $F^{\times}$ we see that we have to integrate on K a function invariant on the left under $(AN) \cap K$. A simple transformation gives then

$$\Psi(s,W_1,W_2,\Phi) = L(2s,\omega)(J + J')$$

where

$$J = \int_R \pi_1\left[w\begin{pmatrix}1 & x\\0 & 1\end{pmatrix}\right]\varphi_1(a)\pi_2\left[w\begin{pmatrix}1 & x\\0 & 1\end{pmatrix}\right]\varphi_2(-a)\,|a|^{s-1}\,d^\times a dx$$

and

$$J' = \int_{\varpi R} \pi_1\left[w\begin{pmatrix}1 & x\\0 & 1\end{pmatrix}w\right]\varphi_1(a)\pi_2\left[w\begin{pmatrix}1 & x\\0 & 1\end{pmatrix}w\right]\varphi_2(-a)\,|a|^{s-1}\,d^\times a dx \ .$$

In J we recognize the integral of Lemma 15.15.1. We may therefore use this lemma and also the formula (2.11.ii)of [1] which gives

$$\sum_{\chi} C_1(-a-b,\chi^{-1})\eta(\chi\mu_2,\varpi^{-a})\eta(\chi\nu_2,\varpi^{-b}) = \omega_1(-\varpi^{-a})\delta_{a,b}$$
$$- (1-q^{-1})^{-1}\omega_1(\varpi)C_1(-a-1,\mu_2)C_1(-b-1,\nu_2)$$
$$- \sum_{r\geq -2}{}' \omega_1(\varpi^{-r})C_1(r-a,\mu_2)C_1(r-b,\nu_2) \ .$$

Now the product

$$C_1(n,\mu_2)C_1(m,\nu_2)$$

vanishes unless $n = m = -c$. Then

$$C_1(-c,\mu_2)C_1(-c,\nu_2) = \omega_1(-\varpi^{-c}) \ .$$

The above expression is then zero unless $a = b$. For $1 \leq a = b < c - 1$ all terms cancel and the expression vanishes. For

$$a = b = c - 1$$

its value is

$$-q^{-1}(1-q^{-1})^{-1}\omega_1(-\varpi^{1-c}) \ .$$

For

$$c \leq a = b \leq 2u$$

its value is

$$\omega_1(-\varpi^{-a}) \ .$$

Setting

$$X = \omega(\varpi^{-1})q^{2s-1} = q^{2s+\sigma-1}$$

we find

$$J = \epsilon(1,\pi_1)\epsilon(1,\pi_2)\omega_1(-1)q^{2u(1-s)} \times (1-X)^{-1}[(1-q^{-1}X^{-1})X^c - (1-q^{-1})X^{2u+1}].$$

We pass to the computation of J'. Both $\pi_1(w)\varphi_1$ and $\pi_2(w)\varphi_2$ are invariant under

$$\begin{pmatrix} 1 & x \\ 0 & 1 \end{pmatrix} \text{ for } x \text{ in } R .$$

So we find

$$J' = q^{-1}\omega(-1)\int\varphi_1(a)\varphi_2(-a)\,|a|^{s-1}d^\times a = \epsilon(1,\pi_1)\epsilon(1,\pi_2)\omega_1(-1)q^{2u(1-s)-1}X^{2u} .$$

Finally, we obtain for Ψ the expression

$$\Psi(s,W_1,W_2,\Phi) = \epsilon(1,\pi_1)\epsilon(1,\pi_2)q^{2u(1-s)}\omega_1(-1)(1-X)^{-1}(X^c-X^{2u+1}) .$$

Changing s in $1-s-\sigma$ we get

$$\tilde{\Psi}(1-s,W_1,W_2,\hat{\Phi}) = \epsilon(1,\pi_1)\epsilon(1,\pi_2)q^{2u(s+\sigma)}\omega_1(-1)(1-X^{-1})(X^{-c}-X^{-2u-1}) .$$

As $\omega_1\mu_2\nu_2$ is unramified, we easily find

$$\epsilon(s,\pi_1 \otimes \mu_2)\epsilon(s,\pi_1 \otimes \nu_2) = \omega_1(-1)X^{-c} .$$

It is now a simple matter to check that

$$\Psi(s,W_1,W_2,\Phi)\epsilon(s,\pi_1 \otimes \mu_2)\epsilon(s,\pi_1 \otimes \nu_2) = \omega_2(-1)\tilde{\Psi}(1-s,W_1,W_2,\hat{\Phi}) .$$

This completes the proof of (15.2).

§16. Explicit computations (continued)

We keep the notations of the previous paragraph. Therefore π_i, $i = 1,2$ are two irreducible admissible representations. We assume that they are not one dimensional. We denote by ω_i the quasi-character of $F^\times$ defined by

$$\pi_i\begin{pmatrix} a & 0 \\ 0 & a \end{pmatrix} = \omega_i(a)1 \quad \text{for all } a \text{ in } F^\times .$$

We set

$$\omega = \omega_1\omega_2 .$$

In this paragraph we shall prove the following theorem.

Theorem 16.1: For $i = 1,2$ let (μ_i,ν_i) a pair of quasi-characters of $F^\times$ such that $\omega_i = \mu_i\nu_i$. Then there is an integer m so that if χ is a quasi-character of order $u \geq m$ then

$$L(s,\pi_1 \times (\pi_2 \otimes \chi)) = L(s,\tilde{\pi}_1 \times (\tilde{\pi}_2 \otimes \chi^{-1})) = 1$$

and

$$\epsilon(s,\pi_1\times(\pi_2\otimes\chi),\psi) = \epsilon(s,\mu_1\mu_2\chi,\psi)\epsilon(s,\mu_1\nu_2\chi,\psi)\epsilon(s,\nu_1\mu_2\chi,\psi)\epsilon(s,\nu_1\nu_2\chi,\psi) .$$

First suppose that $\pi_2 = \pi(\mu_3,\nu_3)$ where μ_3 and ν_3 are such that $\mu_3\nu_3 = \omega_2$. Then for all quasi-characters χ we have by (15.1)

$$L(s,\pi_1\times(\pi_2 \otimes \chi)) = L(s,\pi_1 \otimes \mu_3\chi)L(s,\pi_1 \otimes \nu_3\chi) ,$$

$$L(s,\tilde{\pi}_1\times(\tilde{\pi}_2 \otimes \chi^{-1})) = L(s,\tilde{\pi}_1 \otimes \mu_3^{-1}\chi^{-1})L(s,\tilde{\pi}_1 \otimes \nu_3^{-1}\chi^{-1}) ,$$

$$\epsilon(s,\pi_1\times(\pi_2 \otimes \chi),\psi) = \epsilon(s,\pi_1 \otimes \mu_3\chi,\psi)\epsilon(s,\pi_1 \otimes \nu_3\chi,\psi) .$$

Then the theorem follows at once from Proposition 3.8 in [1].

If $\pi_2 = \sigma(\mu_3,\nu_3)$ where $\mu_3\nu_3 = \omega_2$ and $\mu_3\nu_3^{-1} = \alpha$ a similar argument can be given. As π_1 and π_2 play the same role, we may assume that π_1 and π_2 are absolutely cuspidal. We claim then that it is enough to prove the following proposition.

Proposition 16.2: Assume that π_i , $i = 1,2$ is absolutely cuspidal. Let (μ_3,ν_3) be a pair of quasi-characters such that $\mu_3\nu_3 = \omega_2$ and $\mu_3\nu_3^{-1} \neq \alpha,\alpha^{-1}$. There is an integer m such that if χ is a quasi-character of order $u \geq m$ then

$$\epsilon'(s,\pi_1\times(\pi_2\otimes\chi),\psi) = \epsilon'(s,\pi_1\times(\pi_3\otimes\chi),\psi) ,$$

where $\pi_3 = \pi(\mu_3,\nu_3)$.

Taking the proposition for granted at the moment, let us prove the theorem. Since the theorem is true for the pair (π_1,π_3) all we have to prove is the assertion on the factors L . Applying the theorem to the pair (π_1,π_3) we see that, if the order of χ is large enough,

$$\epsilon'(s,\pi_1\times(\pi_3\otimes\chi),\psi) = \epsilon(s,\pi_1\times(\pi_3\otimes\chi),\psi)$$

is a constant times a power of $X = q^{-s}$. By the proposition we may further assume that the same is true of

$$\epsilon'(s,\pi_1\times(\pi_2\otimes\chi),\psi) .$$

Then the quotient

$$L(s,\pi_1\times(\pi_2\otimes\chi))/L(1-s,\tilde{\pi}_1\times(\tilde{\pi}_2\otimes\chi^{-1}))$$

is also a power of q^{-s} times a constant.

Suppose that

$$L(s,\pi_1\times(\pi_2\otimes\chi))^{-1}$$

is not identically one. Then it is divisible by a factor

$$1 - aX .$$

It follows that

$$L(s,\tilde{\pi}_1\times(\tilde{\pi}_2\otimes\chi^{-1}))^{-1}$$

is divisible by

$$1 - a'X$$

where

$$aa' = q \quad .$$

Since we assume π_1 and π_2 absolutely cuspidal, their Kirillov model is the space $\mathcal{S}(F^{\times})$. Hence, for W_1 in $\mathcal{W}(\pi_1,\psi)$ and W_2 in $\mathcal{W}(\pi_2 \otimes \chi,\psi)$, we see that the only poles of $\Psi(s,W_1,W_2,\Phi)$ are poles of

$$z(\alpha^{2s}\omega^2,k.\Phi) \ , \quad k \in K \ .$$

Hence the quotient

$$L(s,\pi_1 \times (\pi_2 \otimes \chi))/L(2s,\omega\chi^2)$$

is a polynomial in $X = q^{-s}$. So we find

$$L(2s,\omega\chi^2) = (1-aX)(1+aX) \quad .$$

In the same manner

$$L(2s,\omega^{-1}\chi^{-2}) = (1-a'X)(1+a'X) \ .$$

Hence

$$(aa')^2 = 1$$

which is a contradiction. Therefore

$$L(s,\pi_1 \times (\pi_2 \otimes \chi)) = 1 \quad .$$

Similarly

$$L(s,\tilde{\pi}_1 \times (\tilde{\pi}_2 \otimes \chi^{-1})) = 1 \ .$$

So it is enough to prove Proposition 16.2. We may assume that ψ has order zero. Then we will write $\epsilon(s,\pi)$ for $\epsilon(s,\pi,\psi)$ ... etc. We choose m strictly larger than the orders of μ_1 , ν_1 , μ_3 , ν_3 . We may also take m so large that if χ has order $u \geq m$ then

$$L(s,\pi_3 \otimes \chi) = L(s,\tilde{\pi}_3 \otimes \chi^{-1}) = 1$$

$$\epsilon(s,\pi_2 \otimes \chi) = \epsilon(s,\pi_3 \otimes \chi) = \epsilon(s,\mu_3\chi)\epsilon(s,\nu_3,\chi) \ ,$$

$$\epsilon(s,\pi_1 \otimes \chi) = \epsilon(s,\mu_1\chi)\epsilon(s,\nu_1\chi) \quad .$$

If we introduce for the representation π_i , $i = 1,2,3$ the quantities

$C_i(n,\chi)$ (which were denoted $C_n(\chi)$ in [1]) we see that if χ is a character of $R^\times$ or order $u \geq m$ then

(16.2.1) $$C_2(n,\chi) = C_3(n,\chi) \quad \text{for all } n \text{ in } \underline{Z} .$$

More precisely these quantities vanish unless $n = -2u$. For $n = -2u$ their common value is

$$(1-q^{-1})^2 q^u \omega_2(\varpi^{-u})\eta(\mu_3\chi,\varpi^{-u})\eta(\nu_3\chi,\varpi^{-u}) .$$

Similarly, if the order of χ is larger than or equal to m , we find that $C_1(n,\chi)$ vanishes unless $n = -2u$ and then takes the value

$$(1-q^{-1})^2 q^u \omega_1(\varpi^{-u})\eta(\mu_1\chi,\varpi^{-u})\eta(\nu_1\chi,\varpi^{-u}) .$$

Let φ_1 be the element of $\mathcal{K}(\pi_1,\psi)$ defined by the conditions

$$\varphi_1(\epsilon a) = \omega_1(\epsilon)\varphi_1(a) \quad \text{for } \epsilon \text{ in } R^\times ,$$

$$\int \varphi_1(a)\,|a|^{s-\frac{1}{2}}\omega_1^{-1}(a)d^\times a = q^{2us}\epsilon(1,\pi_1) .$$

Then

$$\int \pi_1(w)\varphi_1(a)\,|a|^{s-\frac{1}{2}}d^\times a = \omega_1(-1)q^{(c-2u)(s-1)}$$

where $c \geq 2$. We may assume $2m > c$ so that the support of $\pi_1(w)\varphi_1$ is contained in $R - \{0\}$.

For $i = 2,3$ let φ_i be the element of $\mathcal{K}(\pi_i \otimes \chi,\psi)$ defined by the conditions

$$\varphi_i(a\epsilon) = \omega_2(\epsilon)\chi^2(\epsilon)\varphi_i(a) \quad \text{for } \epsilon \text{ in } R^\times ,$$

$$\int \varphi_i(a)\,|a|^{s-\frac{1}{2}}\omega_2(a)^{-1}\chi(a)^{-2}d^\times a = q^{2us}\epsilon(1,\pi_i \otimes \chi) .$$

We assume that the order of χ is $u \geq m$. Then

$$\int \pi_i(w)\varphi_i(a)\,|a|^{s-\frac{1}{2}}d^\times a = \omega_2(-1) .$$

The reader will note that φ_2 and φ_3 are actually the same element of $\mathcal{S}(F^\times)$.

We shall make use of the following lemma.

Lemma 16.3: For $i = 2,3$, let J_i be the integral

$$\int_{F^\times \times R} \pi_1\left[w\begin{pmatrix}1 & x\\ 0 & 1\end{pmatrix}\right]\varphi_1(a)\pi_2\left[w\begin{pmatrix}1 & x\\ 0 & 1\end{pmatrix}\right]\varphi_i(-a)\,|a|^{s-1}d^\times a\,dx .$$

If $u \geq m$, its value is the same for $i = 2$ and $i = 3$.

We assume $u \geq m$. Furthermore, we suppose that $\chi(\varpi) = 1$ (there is no harm in that). Then, as in Lemma 15.15.1, we find that

$$J_i\,\varepsilon(1,\pi_1)^{-1}\varepsilon(1,\pi_i\otimes\chi)^{-1} = \sum_{n,0\leq O(\zeta)\leq 2u} C_1(n-2u,\zeta^{-1})C_i(n-2u,\zeta\chi)q^{-n(s-1)}$$

$$= \sum_{n,0\leq O(\zeta)\leq 2u} C_1(n-2u,\zeta^{-1}\chi)C_i(n-2u,\zeta)q^{-n(s-1)} .$$

This in turn can be written as the sum of

$$(16.3.1)\qquad \sum_{n,0\leq O(\zeta)\leq u,O(\zeta^{-1}\chi)=u} C_1(n-2u,\zeta^{-1}\chi)C_i(n-2u,\zeta)q^{-n(s-1)}$$

and

$$(16.3.2)\qquad \sum_{n,u\leq O(\zeta)\leq 2u,O(\zeta^{-1}\chi)\neq u} C_1(n-2u,\zeta^{-1}\chi)C_i(n-2u,\zeta)q^{-n(s-1)} .$$

Since in (16.3.2) the order of ζ is larger than or equal to m , it is clear that (16.3.2) has the same value for $i = 2$ and 3 . It remains to see that the same is true of (16.3.1). In (16.3.1) $C_1(n,\zeta^{-1}\chi)$ vanishes unless $n = -2u$ and has then a known value. Replacing, we find that (16.3.1) is equal to

$$(1-q^{-1})^2q^u\omega_1(\varpi^{-u})\sum \eta(\mu_1\chi\zeta^{-1},\varpi^{-u})\eta(\nu_1\chi\zeta^{-1},\varpi^{-u})C_i(-2u,\zeta) ,$$

the sum being extended to all characters ζ of $R^\times$ satisfying the conditions

$$0 \leq O(\zeta) \leq u \ , \ O(\zeta^{-1}\chi) = u \ .$$

We may as well sum for all characters ζ of $R^\times$, because the extra terms that we add by so doing are, in fact, zero.

If $\omega\zeta^2$ is ramified of order $v > 0$, we use formula (2.11.i) of [1] to find that (16.3.1) is equal to

$$(1-q^{-1})^2 q^u \omega_1(-\varpi^{-u})\omega_2(-\varpi^v)\eta(\omega^{-1}\chi^{-2},\varpi^{-v})C_i(-u-v,\mu_1\chi)C_i(-u-v,\nu_1\chi) \ .$$

Since $\mu_1\chi$ and $\nu_1\chi$ have order $u \geq m$ we see that this expression has the same value for $i = 2$ or 3. (Cf. 16.2.1).

If on the contrary, $\omega\chi^2$ is unramified we use formula (2.11.ii) of [1] to find that (16.3.1) is proportional to

$$\omega_2(-\varpi^{-u}) + (q^{-1}-1)^{-1}\omega_2(\varpi)C_i(-u-1,\mu_1\chi)C_i(-u-1,\nu_1\chi)$$

$$- \sum_{r\geq -2} \omega_2(\varpi^{-r})C_i(r-u,\mu_1\chi)C_i(r-u,\nu_1\chi)$$

$$= \omega_2(-\varpi^{-u}) - \omega_2(\varpi^u)C_i(-2u,\mu_1\chi)C_i(-2u,\nu_1\chi) = 0 \ .$$

The lemma is therefore proved.

We shall need also the following precision.

Lemma 16.4: *Assume that $\omega\chi^2$ is unramified. Then, for $i = 2,3$, the integral J_i is a power series in q^{-s} which does not contain q^{2us}.*

Using the previous results we see that the coefficient of q^{2us} in J_i is proportional to

$$\sum C_1(-4u,\zeta^{-1}\chi)C_i(-4u,\zeta)$$

where we sum for all characters ζ of order $-2u$. This in turn is proportional to

$$\sum c_1(-4u,\zeta^{-1}\chi)\eta(\chi\mu_3\zeta,\varpi^{-2u})\eta(\chi\nu_3\zeta,\varpi^{-2u})$$

where we sum for all characters ζ of order $-2u$. It amounts to the same to sum for all characters. Using (2.11.ii) we find that this expression vanishes. Hence the lemma.

Let us prove Proposition 16.2. Let W_1 be the element of $\mathfrak{W}(\pi_1,\psi)$ corresponding to φ_1 and, for $i = 2,3$, let W_i be the element of $\mathfrak{W}(\pi_i \otimes \chi,\psi)$ corresponding to φ_i.

Assume that $\omega\chi^2$ is ramified of order $v > 0$. Take Φ in $\mathcal{S}(F^2)$ defined by

$$\Phi(x,y) = \Phi_{\omega\chi^2}(x)\Phi_0(y) \quad .$$

Computing as in (15.15) we find that

$$\Psi(s,W_1,W_i,\Phi) = J_i$$

and

$$\begin{aligned}
&\tilde{\Psi}(1-s,W_1,W_i,\hat{\Phi}) \\
= \quad &\int_{F^\times\chi\varpi^v R} \pi_1\left[w\begin{pmatrix}1 & x\\ 0 & 1\end{pmatrix}w\right]\varphi_1(a)\pi_i\left[w\begin{pmatrix}1 & x\\ 0 & 1\end{pmatrix}w\right]\varphi_i(-a)\,|a|^{-s}\chi^{-2}\omega^{-1}(a)d^\times a \\
= \quad &q^{-v}\omega_1(-1)\int \varphi_1(a)\varphi_2(a)\,|a|^{-s}\chi^{-2}(a)\omega^{-1}(a)d^\times a \\
= \quad &q^{2u(1-s)-v}\omega_1(-1)\epsilon(1,\pi_1)\epsilon(1,\pi_i \otimes \chi) \ .
\end{aligned}$$

So we see that

$$\tilde{\Psi}(1-s,W_1,W_2,\hat{\Phi}) = \tilde{\Psi}(1-s,W_1,W_3,\hat{\Phi}) \neq 0 \quad ,$$

and on the other hand, by Lemma 16.3,

$$\Psi(s,W_1,W_2,\Phi) = \Psi(s,W_1,W_3,\Phi) \ .$$

The proposition follows.

Assume that $\omega\chi^2$ is unramified. Take Φ to be

$$\Phi(x,y) = \Phi_0(x)\Phi_0(y) .$$

Then we find

$$\Psi(s,W_1,W_i,\Phi)L(2s,\chi^2\omega)^{-1}$$

$$= \quad J_i + \int_{\varpi R} \pi_1\left[w\begin{pmatrix}1 & x\\ 0 & 1\end{pmatrix}w\right]\varphi_1(a)\pi_i\left[w\begin{pmatrix}1 & x\\ 0 & 1\end{pmatrix}w\right]\varphi_2(-a)\,|a|^{s-1}d^\times a$$

$$= \quad J_i + \omega_1(-1)q^{-1}\int \varphi_1(a)\varphi_2(a)\,|a|^{s-1}d^\times a$$

$$= \quad J_i + q^{2u(s+\sigma)-1}\epsilon(1,\pi_1)\epsilon(1,\pi_i\otimes\chi)\omega_1(-1) ,$$

where σ is a complex number such that

$$\omega\chi^2 = \alpha^\sigma .$$

This expression has the same value for $i = 2$ or 3. Moreover by Lemma 16.4 it does not vanish identically.

On the other hand,

$$\tilde{\Psi}(1-s,W_1,W_i,\hat{\Phi}) = \Psi(1-s-\sigma,W_1,W_i,\Phi) .$$

So we find that

$$\Psi(s,W_1,W_2,\Phi) = \Psi(s,W_1,W_3,\Phi) \neq 0$$

and

$$\tilde{\Psi}(1-s,W_1,W_2,\hat{\Phi}) = \tilde{\Psi}(1-s,W_1,W_3,\hat{\Phi}) .$$

Again the proposition follows.

Remark: If the characteristic of the ground field F is not 2, there are only a finite number of quadratic characters. It follows that, if the order of χ is sufficiently large, the character $\omega\chi^2$ is ramified. The proof is then a little simpler.

§17. Real case

The real and complex cases are treated by direct verification. For this, we need some simple results of combinatorial analysis that we first discuss.

It will be convenient to adopt the following convention. When we sum with respect to some indexes j,k,... etc. and do not indicate the range in which the indexes have to vary, this range is determined by the condition that all integers which appear in the formula have to be positive or zero. For instance, with this convention, the binomial formula reads simply

$$(X + Y)^n = \sum_i \binom{n}{i} X^{n-i} Y^i .$$

For each integer $n \geq 0$ we set

$$(17.1.1) \qquad (X)_n = X(X+1) \ldots (X+n-1) = \frac{\Gamma(X+n)}{\Gamma(X)} .$$

Hence

$$(17.1.2) \qquad (X)_0 = 1 , \ (X)_{m+n} = (X)_m (X+m)_n ,$$

$$(17.1.3) \qquad (X)_n = (-1)^n (1-X-n)_n .$$

We introduce two linear operators T and Δ on $\underline{C}[X]$. They are defined by

$$(17.1.3) \qquad TP(X) = P(X-1) ,$$

$$(17.1.4) \qquad \Delta P(X) = P(X) - P(X-1) .$$

Then

$$(17.1.4) \qquad \Delta (X)_0 = 0 \ \text{ and } \ \Delta (X)_n = n(X)_{n-1} .$$

As the $(X)_n$ are a linear basis of $\underline{C}[X]$ it follows that $\Delta P = 0$ if and only if P is a constant.

Similarly on $\underline{C}[X,Y]$ we introduce $T_X, T_Y, \Delta_X, \Delta_Y$. For instance

$$\Delta_X P(X,Y) = P(X,Y) - P(X-1,Y) .$$

A number of relations can be proved just by using the fact that

$$\Delta + T = 1 .$$

For instance:

$$(X-n)_p = \sum_i (-1)^i \binom{n}{p-i} \frac{p!}{i!} (X)_{p-i} .$$

The "binomial formula"

$$(17.1.5) \qquad (X+Y)_n = \sum_i \binom{n}{i} (X)_i (Y)_{n-i}$$

is easily proved by induction on n ; assuming the formula of order $n-1$ to be true, we find that the difference between the right and the left-hand side is a polynomial $P(X,Y)$ such that

$$\Delta_X P = 0 , \quad P(0,Y) = 0 .$$

It follows that $P = 0$, hence the formula.

Simple consequences are the following ones:

if

$$\sum c_{i,j} X^i Y^j = (X+Y)^a X^p Y^q$$

then

$$(17.1.6) \qquad \sum c_{i,j} (X)_i (Y)_j = (X+Y+p+q)_a (X)_p (Y)_q ;$$

if $a = 0,1$ and $n \geq 0$, then

$$(17.1.7) \quad (\tfrac{1}{2} - s + \tfrac{1}{2}a)_n = \sum_p (-1)^p 4^{p-n} \frac{(2n+a)!}{(n-p)!(2p+a)!} (s+\tfrac{1}{2}a)_p ;$$

If $p \geq 0$ and $-p \leq n$, then

$$(17.1.8) \qquad (1 - s + \tfrac{1}{2}n)_p = \sum (-1)^i \binom{p}{i} \frac{(n+p)!}{(n+i)!} (s+\tfrac{1}{2}n)_i .$$

[According to our convention, the range of i in the last formula is defined by the inequalities

$$0 \leq i \leq p \;,\; n + i \geq 0 \;.]$$

Now we come back to the problem of the local functional equation. In this paragraph, the ground field is $\underline{R}$. The group K is the group $\underline{O}(2,\underline{R})$. We fix a non-trivial additive character ψ of $\underline{R}$. Then

$$\psi(x) = \exp(2i\pi a x) \;,\; \text{where} \;\; a \neq 0 \;.$$

We define $\mathcal{S}(\underline{R}^2,\psi)$ as the subspace of $\mathcal{S}(R^2)$ whose elements have the form

$$P(x,y)\exp(-\pi|a|(x^2 + y^2)) \;.$$

It is clear that if Φ belongs to $\mathcal{S}(R^2,\psi)$ so does its Fourier transform $\hat{\Phi}$ defined by the formula

$$\hat{\Phi}(x,y) = \int \Phi(u,v)\psi(uy-vx)dudv \;.$$

Let also π_i for $i = 1,2$ to be two irreducible admissible representations of $\mathcal{H}(G,K)$. Let ω_i be the quasi-character of $\underline{R}^\times$ such that

$$\pi_i\begin{pmatrix} a & 0 \\ 0 & a \end{pmatrix} = \omega_i(a) \;.$$

We set $\omega = \omega_1\omega_2$. We assume also that π_i is infinite dimensional. Therefore the space $\mathcal{W}(\pi_i,\psi)$ is defined.

As in §14, for W_i in $\mathcal{W}(\pi_i,\psi)$ and Φ in $\mathcal{S}(\underline{R}^2)$ we define the integrals

$$\Psi(s,W_1,W_2,\Phi) \;,\; \tilde{\Psi}(s,W_1,W_2,\Phi) \;.$$

There is a slight difficulty in defining the representation

$$\pi = \pi_1 \times \pi_2$$

of $\mathcal{H}[GL(2,\underline{R}) \times GL(2,R)]$. For this Hecke algebra is larger than the tensor product of $\mathcal{H}(GL(2,\underline{R}))$ with itself. Actually, to define the representation π, one has to use the enveloping algebra. Since we use π purely as a notational device, we ignore the difficulty.

__Theorem 17.2__: 1) __There is__ s_0 __so that for__ $\operatorname{Re} s > s_0$, W_i __in__ $\mathcal{W}(\pi_i,\psi)$ __and__ Φ __in__ $\mathcal{S}(\underline{R}^2)$ __the integrals__ $\Psi(s,W_1,W_2,\Phi)$ __and__ $\tilde{\Psi}(s,W_1,W_2,\Phi)$ __are absolutely convergent.__

2) __There are two Euler factors__ $L(s,\pi)$ __and__ $L(s,\tilde{\pi})$ __with the following properties.__ __For__ Φ __in__ $\mathcal{S}(R^2,\psi)$ __set__

$$\Psi(s,W_1,W_2,\Phi) = L(s,\pi)\Xi(s,W_1,W_2,\Phi) ,$$

$$\tilde{\Psi}(s,W_1,W_2,\Phi) = L(s,\tilde{\pi})\tilde{\Xi}(s,W_1,W_2,\Phi) .$$

__Then__ $\Xi(s,W_1,W_2,\Phi)$ __and__ $\tilde{\Xi}(s,W_1,W_2,\Phi)$ __have the form__

$$P(s)\,|a|^{-2s}$$

__where__ P __is a polynomial.__

3) __There are families__ W_1^j, W_2^j, Φ^j __so that__

$$\sum_j \Xi(s,W_1^j,W_2^j,\Phi^j) = |a|^{-2s} \quad \left(\text{resp. } \sum_j \tilde{\Xi}(s,W_1^j,W_2^j,\Phi^j) = |a|^{-2s}\right) .$$

4) __There is a factor__ $\epsilon(s,\pi,\psi)$ __which is a constant times a power of__ $|a|^{-s}$ __so that__

$$\tilde{\Xi}(1-s,W_1,W_2,\hat{\Phi}) = \omega_2(-1)\epsilon(s,\pi,\psi)\Xi(s,W_1,W_2,\Phi)$$

__for all__ W_i __in__ $\mathcal{W}(\pi_i,\psi)$ __and__ Φ __in__ $\mathcal{S}(\underline{R}^2,\psi)$.

Formula (14.8.4) is not true for all g, because the spaces $\mathcal{W}(\pi_i,\psi)$ and $\mathcal{S}(\underline{R}^2,\psi)$ are not stable under G. However, they are stable under the enveloping algebra and there is an infinitesimal formula analogous to (14.8.4). It reads

(17.2.5) $\Psi(s,\pi_1(X)W_1,W_2,\Phi) + \Psi(s,W_1,\pi_2(X)W_2,\Phi) + \Psi(s,W_1,W_2,\Phi_1)$

$$= -s\,\mathrm{Tr}X \cdot \Psi(s,W_1,W_2,\Phi) \quad .$$

Here X is an element of the Lie algebra of $G_{\underline{R}}$ (i.e. a 2 by 2 matrix) and Φ_1 is defined by

$$\Phi_1 = \frac{d}{dt} \exp(tx).\Phi\,\Big|_{t=0} \quad .$$

On the other hand (14.8.4) is true for all g in K. Similar remarks apply to the integral $\tilde{\Psi}$.

From this follows, exactly as in [8], Section 8, that if (1) and (2) have been proved, then there are Euler factors satisfying the conditions 2 and 3 (unless Ψ vanishes identically). Moreover, up to multiplication by a constant, they are uniquely determined by these conditions.

Also, if we take for ψ the character defined by

$$\psi(x) = \exp(2i\pi x)$$

then the factor $\epsilon(s,\pi,\psi)$ is actually independent of s.

To give the values of $L(s,\pi)$ and $\epsilon(s,\pi,\psi)$ we have to remind the reader of the classification of the representations of $\mathcal{H}(G,K)$.

First the Weil group $W(\underline{C}/\underline{R}) = W$ is the only non-trivial extension

$$1 \to \underline{C}^{\times} \to W \to \underline{G} \to 1$$

where $\underline{G}$ is the Galois group of $\underline{C}$ over $\underline{R}$, i.e., a group with two elements 1 and τ. One can write the elements of W in the form x or tx where x is in $\underline{C}^{\times}$ and t is an element of W whose class modulo $\underline{C}^{\times}$ is τ. The multiplication is defined by the conditions

$$txt^{-1} = \bar{x} \quad \text{and} \quad t^2 = -1 \ .$$

The homomorphism from W to $\underline{R}^{\times}$ which sends z to $z\bar{z}$ and t to -1

defines an isomorphism of $W/[W,W]$ onto $\underline{R}^{\times}$.

In particular, we may identify the quasi-characters of $\underline{R}^{\times}$ to the one dimensional representations of W . The other irreducible representations of W are two dimensional. They have the form $\mathrm{Ind}(W,\underline{C}^{\times},\chi)$ where χ is a quasi-character of $\underline{C}^{\times}$ such that $\chi^{\tau} \neq \chi$.

To every (semi-simple) finite dimensional representation σ of W we associate a factor $L(s,\sigma)$ and a factor $\epsilon(s,\sigma,\psi)$. They are defined by the following conditions.

If σ is the direct sum of the σ_i then

$$L(s,\sigma) = \prod_i L(s,\sigma_i) \ , \ \epsilon(s,\pi,\psi) = \prod_i \epsilon(s,\sigma_i,\psi) \ .$$

If σ is one dimensional and corresponds to the quasi-character ω of $\underline{R}^{\times}$ then

$$L(s,\sigma) = L(s,\omega) \ , \ \epsilon(s,\sigma,\psi) = \epsilon(s,\omega,\psi) \ .$$

If σ has the form $\mathrm{Ind}(W,\underline{C},\chi)$ where χ is a quasi-character of $\underline{C}^{\times}$ then

$$L(s,\sigma) = L(s,\chi)$$

$$\epsilon(s,\sigma,\psi) = \epsilon(s,\chi,\psi_{\underline{C}})\lambda(\underline{C}/\underline{R},\psi)$$

where $\psi_{\underline{C}}$ is the character of $\underline{C}$ defined by

$$\psi_{\underline{C}} = \psi \circ \mathrm{Tr}$$

and the factor $\lambda(\underline{C}/\underline{R},\psi)$ has been defined in [1] §1. We remind the reader that its value is

$$a|a|^{-1}i$$

if

$$\psi_{\underline{R}}(x) = \psi(x) = e^{2i\pi ax} \ .$$

If σ is a two dimensional representation of W there is a unique irreducible admissible representation π of $\mathcal{H}(G,K)$ such that

$$\pi\begin{pmatrix} a & 0 \\ 0 & a \end{pmatrix} = \omega(a)$$

where ω is the quasi-character of $\underline{R}^{\times}$ corresponding to the one dimensional representation $\det \sigma$ and, for all quasi-characters χ of $\underline{R}^{\times}$,

$$L(s,\pi \otimes \chi) = L(s,\sigma \otimes \chi) ,$$

$$L(s,\tilde{\pi} \otimes \chi^{-1}) = L(s,\tilde{\sigma} \otimes \chi^{-1}) ,$$

$$\epsilon(s,\pi \otimes \chi,\psi) = \epsilon(s,\sigma \otimes \chi,\psi) .$$

The representation π is noted $\pi(\sigma)$. (Cf. [1], §12).

It turns out that the rule which gives the factors of Theorem 17.2 is as simple as possible.

<u>Proposition 17.3</u>: <u>For</u> $i = 1,2$ <u>let</u> ρ_i <u>be a two dimensional representation of</u> W . <u>Let</u> $\pi_i = \pi(\rho_i)$ <u>and</u> $\pi = \pi_1 \times \pi_2$. <u>Then, if</u> π_1 <u>and</u> π_2 <u>are infinite dimensional,</u>

$$L(s,\pi) = L(s,\rho_1 \otimes \rho_2) ,$$

$$L(s.\tilde{\pi}) = L(s,\tilde{\rho}_1 \otimes \tilde{\rho}_2) ,$$

$$\epsilon(s,\pi,\psi) = \epsilon(s,\rho_1 \otimes \rho_2,\psi) .$$

Once the following lemma is granted, the proof is little more than an exercise in combinatorial analysis.

<u>Lemma 17.3.1</u>: <u>Let</u> a,b,c,d <u>be four complex numbers</u>. <u>Assume that there is a path from</u> $-i\infty$ <u>to</u> $+i\infty$ <u>which is so curved that the poles of</u> $\Gamma(c-s)\Gamma(d-s)$ <u>lie on the right of the path and the poles of</u> $\Gamma(s+a)\Gamma(s+b)$ <u>lie on the left</u>. <u>Integrating along this path, we find</u>

$$(2i\pi)^{-1}\int \Gamma(s+a)\Gamma(s+b)\Gamma(c-s)\Gamma(d-s)ds = \frac{\Gamma(a+c)\Gamma(a+d)\Gamma(b+c)\Gamma(b+d)}{\Gamma(a+b+c+d)} .$$

This is Barnes' lemma. For a proof, see for instance Whittaker and Watson's book, [24] (14.52).

For our purposes, it will be convenient to give the following formulation of Barnes' lemma.

Lemma 17.3.2: For $i = 1,2$ let φ_i be a function on $\underline{R}^\times$ such that, for $\mathrm{Re}\, s$ large enough, the integral

$$\int \varphi_i(t)\,|t|^{s-\frac{1}{2}}\, d^\times t$$

is absolutely convergent and equal to

$$G_1(s+\sigma_i)G_1(s+\tau_i) .$$

Assume also that at least one of the functions φ_i is even. Then, for $\mathrm{Re}\, s$ large enough, the integral

$$\int \varphi_1(t)\varphi_2(t)\,|t|^{s-1}\, d^\times t$$

is absolutely convergent and equal to

$$\frac{G_1(s+\sigma_1+\sigma_2)G_1(s+\sigma_1+\tau_2)G(s+\tau_1+\sigma_2)G(s+\sigma_1+\sigma_2)}{G_1(2s+\sigma_1+\tau_1+\sigma_2+\tau_2)} .$$

Here dx is the ordinary Lesbegue measure on the real line. It is self dual with respect to the character $\psi(x) = \exp(2i\pi x)$. Then $d^\times t$ is the measure $|t|^{-1}dt$. The function $G_1(s)$ is defined by

$$G_1(s) = \pi^{-s/2}\,\Gamma(s/2) .$$

Hence

$$G_1(s) = \int \exp(-\pi x^2)\,|x|^s\, d^\times x .$$

We may replace φ_i by $\varphi_1(t)\,|t|^{-\frac{1}{2}}$ and assume that the two functions

φ_i are even. Then they have the integral representation

$$\varphi_i(t) = \int \exp(-\pi(t^2a^2+a^{-2}))\,|a|^{\sigma_i-\tau_i}\, d^\times a\, |t|^{\sigma_i} .$$

It easily follows that for $|t|$ large φ_i is rapidly decreasing and for $|t|$ small slowly increasing. In particular, the integral

$$\int \varphi_1(t)\varphi_2(t)\,|t|^s\, d^\times t$$

is absolutely convergent if $\mathrm{Re}\, s$ is large enough. We may also replace φ_i by the function $\varphi_i(t)|t|^s$ (which amounts to replace σ_i and τ_i by σ_i+s and τ_i+s respectively) and assume that φ_i is integrable and square integrable on the group $\underline{R}^\times$. Now the dual of the group $\underline{R}_+^\times$ is the group $\underline{R}$, the dual of the Haar measure $d^\times t$ being the measure $(2\pi)^{-1}dt$. We have

$$\int_{\underline{R}_+^\times} \varphi_i(t)t^{iu}\, d^\times t = \tfrac{1}{2}\, G_1(iu+\sigma_i)G_1(iu+\tau_i) .$$

Hence, by Fourier theorem,

$$\int_{\underline{R}_+^\times} \varphi_1(t)\varphi_2(t)d^\times t = \int (2\pi)^{-1}4^{-1}\, G_1(iu+\sigma_1)G_1(iu+\tau_1)G(\sigma_2-iu)G(\tau_2-iu)du .$$

Hence

$$\int_{\underline{R}} \varphi_1(t)\varphi_2(t)d^\times t = (2i\pi)^{-1}\int_{-i\infty}^{+i\infty} G_1(2s+\sigma_1)G_1(2s+\tau_1)G(\sigma_2-2s)G(\tau_2-2s)ds .$$

Using Barnes' lemma, we find that the value of this integral is

$$\frac{G_1(\sigma_1+\sigma_2)G_1(\sigma_1+\tau_2)G(\tau_1+\sigma_2)G(\tau_1+\tau_2)}{G_1(\sigma_1+\sigma_2+\tau_1+\tau_2)} .$$

The lemma is proved.

As for the convergence of the integrals Ψ and $\tilde{\Psi}$, we have only to show that if W_i belongs to $\mathfrak{W}(\pi_i,\psi)$ the integral

$$\int \varphi_1(t)\varphi_2(t)\,|t|^{s-1}\, d^\times t ,$$

where

$$\varphi_i(t) = W_i\begin{pmatrix} t & 0 \\ 0 & 1 \end{pmatrix} ,$$

is absolutely convergent if Res is large enough. But the functions φ_i satisfy conditions similar to the conditions in Lemma 17.3.2 and our assertion is therefore obvious.

We need to compute the factors $z(\alpha^{2s}\omega,\Phi)$ and $z(\alpha^{2s}\omega^{-1},\hat{\Phi})$. The following lemma takes care of that. (Here $\psi(x) = e^{2i\pi x}$).

Lemma 17.3.3: Let $n \geq 0$ be an integer. Then every element of $\mathcal{S}(\underline{R}^2,\psi)$ which satisfy

$$k.\Phi = e^{-in\theta}\Phi \quad \text{for} \quad k = \begin{pmatrix} \cos\theta & \sin\theta \\ -\sin\theta & \cos\theta \end{pmatrix}$$

is a linear combination of functions of the form

$$\Phi(x,y) = \exp(-\pi(x^2+y^2))(x+iy)^p(x-iy)^q ,$$

where $p \geq 0$, $q \geq 0$, and $q-p = n$. If Φ is the above function and ω the quasi-character defined by

$$\omega(x) = |x|^\sigma(|x|^{-1}x)^\epsilon \quad \text{where} \quad \epsilon = 0 \text{ or } 1 ,$$

the integrals $z(\alpha^{2s}\omega,\Phi)$ and $z(\alpha^{2s}\omega^{-1},\hat{\Phi})$ vanish unless ϵ and n have the same parity. If it is the case, then

$$z(\alpha^{2s}\omega,\Phi) = (-i)^n G_1(2s+\sigma+n)\pi^{-p}(s+\tfrac{1}{2}\sigma+\tfrac{1}{2}n)_p ,$$

$$z(\alpha^{2-2s}\omega^{-1},\hat{\Phi}) = (i)^n G_1(2-2s-\sigma+n)\pi^{-p}(s+\tfrac{1}{2}\sigma+\tfrac{1}{2}n)_p .$$

The first assertion is left to the reader. The assertions relative to $z(\alpha^{2s}\omega,\Phi)$ are immediate if we take in account the relation

$$G_1(s+2p) = \pi^{-p}G_1(s)(\tfrac{1}{2}s)_p .$$

Now we observe that if the function Φ has the form

$$\Phi(x,y) = e^{-\pi z\bar{z}}\, P(z,\bar{z})$$

where $z = x+iy$ and P is a polynomial, then

$$(\bar{z}.\Phi)^\wedge = \pi^{-1}\frac{\partial}{\partial z}\hat{\Phi}\ ,\ (z.\Phi)^\wedge = \pi^{-1}\frac{\partial}{\partial \bar{z}}\hat{\Phi}\ .$$

It follows that the Fourier transform of the function

$$e^{-\pi z\bar{z}}\ \bar{z}^{n}$$

is the function

$$(-1)^n\ e^{-\pi z\bar{z}}\ \bar{z}^{n}\ ,$$

and that the Fourier transform of the function

$$\Phi = e^{-\pi z\bar{z}}\ z^p\ \bar{z}^q$$

is given by

$$\hat{\Phi}(x,y) = \sum \frac{p!q!(-1)^{q-i}\pi^{-1}}{(p-i)!(q-i)!i!}\ z^{p-i}\ \bar{z}^{q-i}\ e^{-\pi z\bar{z}}\ .$$

So we find that $z(\alpha^{2-2s}\omega^{-1},\hat{\Phi})$ vanishes unless $\epsilon \equiv n \bmod 2$; in that case:

$$z(\alpha^{2-2s}\omega^{-1},\hat{\Phi}) = \sum \frac{p!q!\pi^{-j}(i)^{p+q-2j}}{(p-j)!(q-j)!j!}\quad G_1(2-2s-\sigma+2p+n-2j)$$

$$= (i)^n G_1(2-2s-\sigma+n)\pi^{-p}\sum\frac{p!(n+p)!(-1)^j}{j!(n+j)!(p-j)!}\ (1-s-\tfrac{1}{2}\sigma+\tfrac{1}{2}n)_j\ .$$

Using (17.1.8) we find the required result. This completes the proof of the lemma.

When $\rho_i = \mu_i \oplus \nu_i$ for $i = 1,2$ then

$$\pi_i = \pi(\rho_i) = \pi(\mu_i,\nu_i)$$

and

$$L(s,\rho_i) = L(s,\mu_1\mu_2)L(s,\mu_1\nu_2)L(s,\nu_1\mu_2)L(s,\nu_1\nu_2)$$
$$L(s,\tilde{\rho}_i) = L(s,\mu_1^{-1}\mu_2^{-1})L(s,\mu_1^{-1}\nu_2^{-1})L(s,\nu_1^{-1}\mu_2^{-1})L(s,\nu_1^{-1}\nu_2^{-1})$$
$$\epsilon(s,\rho,\psi) = \epsilon(s,\mu_1\mu_2,\psi)\epsilon(s,\mu_1\nu_2,\psi)\epsilon(s,\nu_1\mu_2,\psi)\epsilon(s,\nu_1\nu_2,\psi)$$
$$= \omega_2(-1)\ .$$

Then (17.2) and (17.3) are, in that case, consequence of the following proposition.

<u>Proposition 17.4</u>: <u>Suppose that</u> $\pi_i = \pi(\rho_i)$. <u>For</u> W_i <u>in</u> $\mathfrak{W}(\pi_i,\psi)$ <u>and</u> Φ <u>in</u> $\mathcal{S}(\underline{R}^2,\psi)$ <u>set</u>

$$\Psi(s,W_1,W_2,\Phi) = \Xi(s,W_1,W_2,\Phi)L(s,\rho_1 \otimes \rho_2)$$

$$\widetilde{\Psi}(s,W_1,W_2,\Phi) = \widetilde{\Xi}(s,W_1,W_2,\Phi)L(s,\widetilde{\rho}_1 \otimes \widetilde{\rho}_2) \quad .$$

<u>Then</u> $\Xi(s,W_1,W_2,\Phi)$ <u>is a polynomial in</u> s <u>and one can find</u> W_1, W_2 <u>and</u> Φ <u>so that</u>

$$\Xi(s,W_1,W_2,\Phi) = 1 \quad .$$

<u>Moreover,</u>

$$\widetilde{\Xi}(1-s,W_1,W_2,\hat{\Phi}) = \Xi(s,W_1,W_2,\Phi) \quad .$$

We have to distinguish the following cases:

(17.4.1) $\quad \mu_1 = \alpha^{\sigma_1} , \nu_1 = \alpha^{\tau_1} , \mu_2 = \alpha^{\sigma_2} , \nu_2 = \alpha^{\tau_2}$;

(17.4.2) $\quad \mu_1 = \alpha^{\sigma_1} , \nu_1 = \alpha^{\tau_1} , \mu_2 = \alpha^{\sigma_2}\xi , \nu_2 = \alpha^{\tau_2}\xi$;

(17.4.3) $\quad \mu_1 = \alpha^{\sigma_1} , \nu_1 = \alpha^{\tau_1} , \mu_2 = \alpha^{\sigma_2}\xi , \nu_2 = \alpha^{\tau_2}$;

(17.4.4) $\quad \mu_1 = \alpha^{\sigma_1}\xi , \nu_1 = \alpha^{\tau_1} , \mu_2 = \alpha^{\sigma_2}\xi , \nu_2 = \alpha^{\tau_2}$.

Here we denote by α the module of $\underline{R}$ and by ξ the character

$$\xi(x) = (|x|^{-1}x) \quad .$$

The fact that we have to study only these cases comes at once from the remark that there is no harm in replacing the pair (π_1,π_2) by the pair $(\pi_1 \otimes \chi, \pi_2 \otimes \chi^{-1})$.

<u>Case 17.4.1</u>: We first remind the reader, if any at this stage, of the recipe to obtain the element of $\mathfrak{W}(\pi_i,\psi)$. For Φ in $\mathcal{S}(\underline{R}^2)$ and g

in $G_{\underline{R}}$ we define $g.\Phi$ by

$$g.\Phi(x,y) = \Phi[(x,y).g] \quad .$$

We also use the partial Fourier transform defined by

$$\tilde{\Phi}(x,y) = \int \Phi(x,u)\exp(2i\pi uy)du$$

to construct a representation r of $G_{\underline{R}}$ on $\mathcal{S}(\underline{R}^2)$. It is defined by

$$[r(g)\Phi]^{\sim} = g.(\tilde{\Phi}) \quad .$$

Then to each Ψ_i in $\mathcal{S}(\underline{R}^2,\psi)$ an element W_i of $\mathcal{W}(\pi_i,\psi)$ is associated by

$$W_i(g) = \int (r(g)\Psi_i)(t,t^{-1})\mu_i\nu_i^{-1}(t)d^{\times}t\ \mu_i(g)\,|\det g|^{\frac{1}{2}} \quad .$$

It amounts to the same to say that for any character χ of $\underline{R}^{\times}$ and any s in $\underline{C}$ with $\operatorname{Re} s$ sufficiently large

$$\int \varphi_i(a)\chi(a)\,|a|^{s-\frac{1}{2}}d^{\times}a = \int \Psi_i(x,y)\mu_i\chi(x)\,|x|^{s}d^{\times}x\ \nu_i\chi(y)\,|y|^{s}d^{\times}y \ ,$$

where φ_i is the function on $\underline{R}^{\times}$ defined by

$$\varphi_i(a) = W_i\begin{pmatrix} a & 0 \\ 0 & 1 \end{pmatrix} .$$

In terms of the Fourier transform $\Phi_i = \tilde{\Psi}_i$, this reads

$$\int \varphi_i(a)\,|a|^{s-\frac{1}{2}}\chi(a)d^{\times}a = \varepsilon'(s,\chi\mu_i,\psi)^{-1}\int \Phi_i(x,y)\mu_i\chi(x)\,|x|^{s-\frac{1}{2}}d^{\times}x \\ \nu_i^{-1}\chi^{-1}(y)\,|y|^{1-s}d^{\times}y \quad .$$

Of course, in the right-hand side, the integral is to be taken in the sense of analytical continuation.

If n is in $\underline{Z}$, there is, up to multiplication by a constant, at most one element W_i of $\mathcal{W}(\pi_i,\psi)$ such that

$$W_i(g.k(\theta)) = e^{in\theta}W_i(g)$$

where $k(\theta)$ is the matrix

$$\begin{pmatrix} \cos\theta & \sin\theta \\ -\sin\theta & \cos\theta \end{pmatrix} .$$

It corresponds to a function Φ_i such that

$$k(\theta).\Phi_i = e^{in\theta}\Phi_i .$$

We take now W_1 to be invariant under $SO(2,\underline{R})$. For W_2 we take a function such that

$$\pi_2(k(\theta))W_2 = e^{i\theta} W_2 \quad \text{where} \quad n \geq 0 .$$

Then we shall prove the assertions of (17.4) for W_1, W_2 and all Φ. Since

$$\Psi(s,W_1,W_2,\Phi) = \Psi[s,\pi_1(\eta)W_1,\pi_2(\eta)W_2,\eta.\Phi]$$

where η is the matrix

$$\begin{pmatrix} -1 & 0 \\ 0 & 1 \end{pmatrix}$$

it will follow that the assertions of (17.4) are true for all W_2 such that

$$\pi_2(k(\theta))W_2 = e^{in\theta}W_2 \quad \text{with} \quad n \text{ in } \underline{Z}$$

and therefore, by linearity, for all W_2. Now using (17.2.5) and the fact that the functions transformed of W_1 span $\mathfrak{W}(\pi_1,\psi)$ we will obtain (17.4) for all choices of W_1 and W_2.

Now W_1 corresponds to

$$\Phi_1(x,y) = \Psi_1(x,y) = \exp(-\pi(x^2+y^2)) .$$

Therefore

$$\int\varphi_1(a)\,|a|^{s-\frac{1}{2}}d^{\times}a = G_1(s+\sigma_1)G_1(s+\tau_1) .$$

As for W_2 it corresponds to

$$\Phi_2(x,y) = \exp(-\pi(x^2+y^2))(x+iy)^n$$

where in fact $n = 2m$. Hence we find

$$\int \varphi_2(a)\,|a|^{s-\frac{1}{2}} d^{\times}a$$

$$= G_1(s+\tau_2)G_1(1-s-\tau_2)^{-1} \sum{}' \binom{2m}{2j} \int \exp(-\pi x^2)x^{2j}\,\mu_2(x)\,|x|^s\,d^{\times}x$$

$$\int \exp(-\pi y^2)(iy)^{2(m-j)}\,\nu_2^{-1}(y)\,|y|^{1-s}\,d^{\times}y \;.$$

or, computing the integrals,

$$\sum \binom{2m}{2j} (-1)^{m-j} G_1(s+\sigma_2+2j)G_1(s+\tau_2)G_1(1-s-\tau_2+2m-2j)G_1(1-s-\tau_2)^{-1} \;.$$

Using the identity

$$G_1(X+2p) = G_1(X)(X/2)_p\,\pi^{-p}$$

we transform this expression into

$$\sum_j \binom{2m}{2j} (-\pi)^{j-m} G_1(s+\sigma_2+2j)G_1(s+\tau_2)\left(\frac{1-s-\tau_2}{2}\right)_{m-j} \;.$$

Using the identity (17.1.7) and again the above identity we find

$$\int \varphi_2(a)\,|a|^{s-\frac{1}{2}} d^{\times}a = \sum_{j,k} \frac{(2m)!\,(-4\pi)^{k+j-m}}{(2j)!\,(2k)!\,(m-j-k)!}\; G_1(s+\sigma_2+2j)G_1(s+\tau_2+2k) \;.$$

With the present choice of W_1 and W_2 we may assume furthermore that Φ satisfies

$$k(\theta).\Phi = e^{-in\theta}\Phi \;.$$

Then, it follows from Lemma 17.3.3 that we may also assume that

$$\Phi(x,y) = \exp(-\pi(x^2+y^2))(x-iy)^n(i)^n \;.$$

Then $\Phi = \hat{\Phi}$ and

$$z(\alpha^{2s}\omega,\Phi) = G_1(2s+\sigma+2m) \;,$$

where $\sigma = \sigma_1 + \tau_1 + \sigma_2 + \tau_2$.

We compute Ψ as an integral on $\underline{R}^{\times} \times SO(2,\underline{R})$. We obtain in that way

$$\Psi(s,W_1,W_2,\Phi) = \int \varphi_1(a)\varphi_2(-a)\,|a|^{s-1}d^\times a\ z(\alpha^{2s}\omega,\Phi)$$

or, using the previous results and Lemma 17.3.2

$$\sum_{j,k} \frac{(2m)!(-4\pi)^{k+j-m}\,G_1(2s+\sigma+2m)}{(2j)!(2k)!(m-j-k)!\,G_1(2s+\sigma+2j+2k}$$

$$\times\ G_1(s+\sigma_1+\sigma_2+2j)G_1(s+\sigma_1+\tau_2+2k)G_1(s+\tau_1+\sigma_2+2j)G_1(s+\tau_1+\tau_2+2k)\ .$$

To find $\Xi(s,W_1,W_2,\Phi)$ we have to divide by

$$L(s,\rho_1\otimes\rho_2) = G_1(s+\sigma_1+\sigma_2)G_1(s+\sigma_1+\tau_2)G_1(s+\tau_1+\sigma_2)G_1(s+\tau_1+\tau_2)\ .$$

Doing so we obtain

$$\Xi(s,W_1,W_2,\Phi) = \pi^{-2m}P(S_1,S_2,T_1,T_2)$$

where we set

$$S_1 = \tfrac{1}{2}(s+\sigma_1+\sigma_2)\ ,\ S_2 = \tfrac{1}{2}(s+\tau_1+\tau_2)$$

$$T_1 = \tfrac{1}{2}(s+\tau_1+\sigma_2)\ ,\ T_2 = \tfrac{1}{2}(s+\sigma_1+\tau_2)$$

and

$$P(S_1,S_2,T_1,T_2) = \sum_{j,k} \frac{(2m)!(-4)^{k+j-m}}{(2j)!(2k)!(m-j-k)!}$$

$$\times\ (S_1+S_2+j+k)_{m-j-k}(S_1)_j(S_2)_k(T_1)_j(T_2)_k\ .$$

We note that

$$S_1+S_2 = T_1+T_2\ .$$

Since P is a polynomial and a constant if $m = 0$ we see that the two first assertions of (17.4) are proved. To check the functional equation we observe that $\Phi = \hat{\Phi}$ and

$$\tilde{\Psi}(1-s,W_1,W_2,\hat{\Phi}) = \Psi(1-s-\sigma,W_1,W_2,\Phi)$$

where, we remind the reader,

$$\sigma = \sigma_1 + \tau_1 + \sigma_2 + \tau_2 .$$

So we find

$$\tilde{\Xi}(1-s,W_1,W_2,\hat{\Phi}) = \pi^{-2m} P(\tfrac{1}{2}-S_2,\tfrac{1}{2}-S_1,\tfrac{1}{2}-T_2,\tfrac{1}{2}-T_1) .$$

So the functional equation to be proved reads

$$\tilde{\Xi}(1-s,W_1,W_2,\hat{\Phi}) = \Xi(s,W_1,W_2,\Phi)$$

and is therefore a consequence of the following lemma.

Lemma 17.4.1: *If* P *is the above polynomial and the variables satisfy*

$$S_1 + S_2 = T_1 + T_2 ,$$

then the polynomial P *satisfies the functional equation*

$$P(S_1,S_2,T_1,T_2) = P(\tfrac{1}{2}-S_2,\tfrac{1}{2}-S_1,\tfrac{1}{2}-T_2,\tfrac{1}{2}-T_1) .$$

Indeed, under the assumptions of the lemma, it is clear that

$$P(S_1,S_2,T_1,T_2) = P(T_1,T_2,S_1,S_2) .$$

On the other hand, we can use the "binomial formula" to write P in the form

$$(-1)^m(2m)! \sum \frac{4^{k+j-m}(-1)^{k+j}}{(2j)!(2k)!p!q!} (S_1)_{j+p}(S_2)_{k+q}(T_1)_j(T_2)_k$$

where we sum now for

$$j + k + p + q = m .$$

We may as well sum for

$$m = \alpha + \beta , \alpha = j + p , \beta = k + q .$$

We get in that way

$$\sum_{\alpha+\beta=m} (S_1)_\alpha(S_2)_\beta \frac{(-1)^m(2m)!}{(2\alpha)!(2\beta)!} \sum_{0\le j\le\alpha} \frac{(-1)^j(2\alpha)!4^{j-\alpha}}{(2j)!(\alpha-j)!}(T_1)_j$$

$$\sum_{0\le k\le\beta} \frac{(-1)^k(2\beta)!4^{k-\beta}}{(2k)!(\beta-k)!}(T_2)_k .$$

Using (17.1.7) we find finally

$$P(S_1,S_2,T_1,T_2) = (-1)^m \sum_{\alpha+\beta=m} \binom{2m}{2\alpha} (S_1)_\alpha (S_2)_\beta (\tfrac{1}{2}-T_1)_\alpha (\tfrac{1}{2}-T_2)_\beta \quad .$$

This shows clearly that P satisfies the identity:

$$P(S_1,S_2,T_1,T_2) = P(\tfrac{1}{2}-T_1,\tfrac{1}{2}-T_2,\tfrac{1}{2}-S_1,\tfrac{1}{2}-S_2) \quad .$$

From this the lemma follows. Hence the Proposition 17.4 is completely proved in case 17.4.1.

Case 17.4.2: The proof is similar. We may assume W_1 and Φ to be as in the previous case and W_2 to correspond to

$$\Phi_2(x,y) = \exp(-\pi(x^2+y^2))(x+iy)^n \ ,$$

where $n = 2m \geq 0$. Then the integral

$$\int \varphi_2(a)|a|^{s-\frac{1}{2}} d^\times a$$

vanishes if $m = 0$ and, if $m > 0$, is equal to

$$\sum \frac{(-4\pi)^{j+k-m+1}(2m)!}{(2j+1)!(2k+1)!(m-j-k-1)!} G_1(s+\sigma_2+2j+1)G_1(s+\tau_2+2k+1) \quad .$$

Finally, we find

$$\Xi(s,W_1,W_2,\Phi) = \pi^{2-2m} P(S_1,S_2,T_1,T_2)$$

where S_1,S_2,T_1,T_2 have the same significance as before and P is the polynomial

$$\sum \frac{(-4)^{j+k-m+1}(2m)!}{(2j+1)!(2k+1)!(m-j-k-1)!}$$

$$\times\ (S_1+S_2+j+k+1)_{m-j-k-1}(S_1+\tfrac{1}{2})_j(S_2+\tfrac{1}{2})_k(T_1+\tfrac{1}{2})_j(T_2+\tfrac{1}{2})_k \quad .$$

An alternate expression for P is

$$\sum_{\alpha+\beta=m-1} \frac{(2m)!(-1)^{m-1}}{(2\alpha+1)!(2\beta+1)!}(S_1+\tfrac{1}{2})_\alpha(S_2+\tfrac{1}{2})_\beta(\tfrac{1}{2}-T_1+\tfrac{1}{2})_\alpha(\tfrac{1}{2}-T_2+\tfrac{1}{2})_\beta \quad .$$

The proof is then the same as in the previous case.

Case 17.4.3: We take W_1 as in (17.4.1), $n = 2m + 1$ and W_2 corresponding to

$$\Phi_2(x,y) = \exp(-\pi(x^2+y^2))(x+iy)^n ,$$

we find

$$\int \varphi_2(a)\,|a|^{s-\frac{1}{2}}d^\times a$$

$$= \sum_{j,k} \frac{(-4\pi)^{j+k-m}(2m+1)!}{(2j+1)!(2k)!(m-j-k)!} G_1(s+\sigma_2+2j+1)G_1(s+\tau_2+2k)$$

and

$$\int \varphi_2(a)\,|a|^{s-\frac{1}{2}}\xi(a)d^\times a$$

$$= \sum_{j,k} \frac{(-4\pi)^{j+k-m}(2m+1)!}{(2j)!(2k+1)!(m-j-k)!} G_1(s+\sigma_2+2j)G_1(s+\tau_2+2k+1) \quad .$$

On the other hand, we take Φ to be

$$\Phi(x,y) = \exp(-\pi(x^2+y^2))(x-iy)^{2m+1}(i)^{2m+1} \quad .$$

Then we get

$$\Xi(s,W_1,W_2,\Phi) = \pi^{-2m}P(S_1,S_2,T_1,T_2)$$

where P is the polynomial

$$\sum_{j,k} \frac{(2m+1)!(-4)^{k+j-m}}{(2j+1)!(2k)!(m-j-k)!}$$

$$\times\ (S_1+S_2+j+k+\tfrac{1}{2})_{m-j-k}(S_1+\tfrac{1}{2})_j(S_2)_k(T_1+\tfrac{1}{2})_j(T_2)_k \quad .$$

Similarly we find

$$\tilde{\Xi}(1-s,W_1,W_2,\hat{\Phi}) = \pi^{-2m}P(\tfrac{1}{2}-S_1,\tfrac{1}{2}-S_2,\tfrac{1}{2}-T_1,\tfrac{1}{2}-T_2) \quad .$$

An alternate expression for P is

$$\sum_{\alpha+\beta=m} \frac{(2m)!(-1)^m}{(2\alpha+1)!(2\beta)!}(S_1+\tfrac{1}{2})_\alpha(S_2)_\beta(\tfrac{1}{2}-T_1+\tfrac{1}{2})_\alpha(\tfrac{1}{2}-T_2)_\beta \quad .$$

The proof is then the same as before.

Case 17.4.4: For $i = 1,2$ one can find a basis

$$W_i^{2j+1}, \quad j \in \underline{Z},$$

of $\mathfrak{W}(\pi_i,\psi)$ satisfying the following conditions for all j in $\underline{Z}$

$$\pi_i(k(\theta))W_i^{2j+1} = e^{(2j+1)\theta}W_i^{2j+1} \quad ;$$

one can find X in the complex Lie algebra of $SL(2,\underline{R})$ such that

$$\pi_i(X)W_i^{2j+1} = W_i^{2j+3} \quad \text{for } j \geq 0$$

$$\pi_i(X)W_i^{2j+1} = \lambda_i^j W_i^{2j+3} \quad \text{for } j \text{ in } \underline{Z} .$$

(It may happen that λ_i^j vanishes for some $i < 0$). It will suffice to prove the following lemma.

Lemma 17.4.4.1: The assertions of Proposition 17.4 are true when (W_1,W_2) is one of the pairs

$$(W_1^{-1},W_2^1) \text{ or } (W_1^1,W_2^{2j+1}) \text{ where } j \geq 0 .$$

Taking the lemma for granted at the moment, we show that (17.4) is true. First, we show that the assertions of (17.4) are true for the pair

$$(W_1^{-1},W_2^{2j+1}) \quad \text{for } j \geq 0 .$$

For $j = 0$ this follows from the lemma. So we may assume $j > 0$ and (17.4) true for the pair

$$(W_1^{-1},W_2^{2j-1}) \quad .$$

By (17.2.5) for all Φ in $\mathcal{S}(\underline{R}^2,\psi)$ we have a relation of the type

$$\Psi(s,W_1^{-1},W_2^{2j+1},\Phi) + \lambda_1^{-1}\Psi(s,W_1^1,W_2^{2j-1},\Phi) + \Psi(s,W_1^{-1},W_2^{2j-1},\Phi_1) = 0 \quad .$$

There is a similar relation for $\tilde{\Psi}$. It follows that (17.4) is true for the pair

$$(W_1^{-1}, W_2^{2j+1}) \quad .$$

Since $\pi_i(\eta)W_i^{2j+1}$ is proportional to W_i^{-2j-1} we find that the assertions of (17.4) are true for all pairs

$$(W_1^{1}, W_2^{2j+1}) \;, \quad j \in \underline{Z} \quad .$$

By linearity, they are true for all pairs

$$(W_1^1, W_2) \quad \text{with} \quad W_2 \quad \text{in} \quad \mathfrak{W}(\pi_2, \psi) \; .$$

Since the functions transformed of W_1^1 under the enveloping algebra and K span $\mathfrak{W}(\pi_1, \psi)$ the proposition follows.

We check the assertions of (17.4) for the pair (W_1^{-1}, W_2^1) . We may replace W_1^{-1} by the function W_1 corresponding to

$$\Phi_1(x,y) = e^{-\pi(x^2+y^2)}(x-iy) \; ,$$

and W_2^1 by the function corresponding to

$$\Phi_2(x,y) = \exp(-\pi(x^2+y^2))(x+iy) \; .$$

Also, it is enough to take

$$\Phi(x,y) = \exp(-\pi(x^2+y^2)) \; .$$

A simple computation gives then

$$\Xi(s, W_1, W_2, \Phi) = 1 \; .$$

Since $\hat{\Phi} = \Phi$ the functional equation to be proved reads

$$\Xi(s, W_1, W_2, \Phi) = \Xi(1-s-\sigma, W_1, W_2, \Phi)$$

which is certainly verified.

We pass to the pair (W_1^1, W_2^{2m-1}) with $m \geq 1$. We replace W_1^1 by the function corresponding to

$$\Phi_1(x,y) = \exp(-\pi(x^2+y^2))(x+iy)$$

and W_2^{2m-1} by the function corresponding to

$$\Phi_2(x,y) = \exp(-\pi(x^2+y^2))(x+iy)^{2m-1} .$$

If, as above, we set

$$\varphi_i(a) = W_i\begin{pmatrix} a & 0 \\ 0 & 1 \end{pmatrix} ,$$

we find (ξ denoting the sign character)

$$\int \varphi_1(a)|a|^{s-\frac{1}{2}}d^{\times}a = G_1(s+\sigma_1+1)G_1(s+\tau_1) ,$$

$$\int \varphi_1(a)\xi(a)|a|^{s-\frac{1}{2}}d^{\times}a = G_1(s+\sigma_1)G_1(s+\tau_1+1) ,$$

and

$$\int \varphi_2(a)|a|^{s-\frac{1}{2}}d^{\times}a = \sum \frac{(-4\pi)^{j+k+1-m}(2m-1)!}{(2j+1)!(2k)!(m-1-j-k)!} G_1(s+\sigma_2+2j+1)G_1(s+\tau_2+2k) ,$$

$$\int \varphi_2(a)\xi(a)|a|^{s-\frac{1}{2}}d^{\times}a = \sum \frac{(-4\pi)^{j+k+1-m}(2m-1)!}{(2j)(2k+1)!(m-1-j-k)!} G_1(s+\sigma_2+2j)G_1(s+\tau_2+2k+1).$$

On the other hand we may assume

$$\Phi(x,y) = (i)^{2m}\exp(-\pi(x^2+y^2))(x-iy)^{2m} .$$

Then we get

$$\Psi(s,W_1,W_2,\Phi) = G_1(2s+\sigma+2m)\int \varphi_1(a)\varphi_2(-a)|a|^{s-\frac{1}{2}}d^{\times}a .$$

Using Lemma 17.3.2 we find

$$\Psi(s,W_1,W_2,\Phi) = \pi^{1-2m}[P(S_1,S_2,T_1,T_2) - P(S_2,S_1,T_2,T_1)] ;$$

where S_i and T_i have the same meaning as before and P is the following polynomial:

$$P(S_1,S_2,T_1,T_2) = \sum \frac{(-4)^{j+k+1-m}(2m-1)!}{(2j+1)!(2k)!(m-1-j-k)!}(S_1+S_2+j+k+1)_{m-j-k-1}$$

$$\times (S_1)_{j+1}(S_2)_k(T_1+\tfrac{1}{2})_j(T_2+\tfrac{1}{2})_k .$$

The only thing to be proved is the functional equation. Since

$$\tilde{\Xi}(1-s,W_1,W_2,\hat{\Phi}) = \Xi(1-s-\sigma,W_1,W_2,\Phi) ,$$

it is a consequence of the following lemma.

Lemma 17.4.4.2: If P is the above polynomial and the variables satisfy the relation

$$S_1 + S_2 = T_1 + T_2$$

then P satisfies the identity

$$P(S_1,S_2,T_1,T_2) - P(S_2,S_1,T_2,T_1) = P(\tfrac{1}{2}-S_2,\tfrac{1}{2}-S_1,\tfrac{1}{2}-T_2,\tfrac{1}{2}-T_1)$$
$$- P(\tfrac{1}{2}-S_1,\tfrac{1}{2}-S_2,\tfrac{1}{2}-T_1,\tfrac{1}{2}-T_2) .$$

Using the same method as before, we find for P the alternate expression

$$(17.4.4.3) \qquad \sum_{\alpha+\beta=m-1} \frac{(-1)^{m-1}(2m-1)!}{(2\alpha+1)!(2\beta)!}(S_1)_{\alpha+1}(S_2)_\beta(1-T_1)_\alpha(-T_2)_\beta .$$

On the other hand, if we assume $S_1+S_2 = T_1+T_2$, we find also

$$(17.4.4.4) \qquad \sum_{\alpha+\beta=m-1} \frac{(-1)^{m-1}(2m-1)!}{(2\alpha+1)!(2\beta)!}(S_1)(T_1+\tfrac{1}{2})_\alpha(T_2+\tfrac{1}{2})_\beta(\tfrac{1}{2}-S_1)_\alpha(\tfrac{1}{2}-S_2)_\beta .$$

Using (17.4.4.4) we find

$$P(\tfrac{1}{2}-S_2,\tfrac{1}{2}-S_1,\tfrac{1}{2}-T_2,\tfrac{1}{2}-T_1) - P(\tfrac{1}{2}-S_1,\tfrac{1}{2}-S_2,\tfrac{1}{2}-T_1,\tfrac{1}{2}-T_2)$$

$$= \sum \frac{(-1)^{m-1}(2m-1)!}{(2\alpha)!(2\beta+1)!}(\tfrac{1}{2}-S_2)(1-T_1)_\alpha(1-T_2)_\beta(S_1)_\alpha(S_2)_\beta$$

$$- \sum \frac{(-1)^{m-1}(2m-1)!}{(2\alpha+1)!(2\beta)!}(\tfrac{1}{2}-S_1)(1-T_1)_\alpha(1-T_2)_\beta(S_1)_\alpha(S_2)_\beta .$$

If in the above expression we use the relations

$$(S_1)_\alpha(\tfrac{1}{2}-S_1) = \tfrac{1}{2}(2\alpha+1)(S_1)_\alpha - (S_1)_{\alpha+1} ,$$

$$(S_2)_\beta(\tfrac{1}{2}-S_2) = -(2\beta+1)(S_2)_\beta - (S_2)_{\beta+1} ,$$

we find that the sum

$$\frac{1}{2}\sum \frac{(-1)^{m-1}(2m-1)!}{(2\alpha)!(2\beta)!}(1-T_1)_\alpha(1-T_2)_\beta(S_1)_\alpha(S_2)_\beta$$

appears once with the plus sign and once with the minus sign. So it cancels and we are left with

$$\sum_{\alpha+\beta=m-1} \frac{(-1)^{m-1}(2m-1)!}{(2\alpha+1)!(2\beta)!}(1-T_1)_\alpha(1-T_2)_\beta(S_1)_{\alpha+1}(S_2)_\beta$$

$$- \sum_{\alpha+\beta=m-1} \frac{(-1)^{m-1}(2m-1)!}{(2\alpha)!(2\beta+1)!}(1-T_1)_\alpha(1-T_2)_\beta(S_1)_\alpha(S_2)_{\beta+1} \quad .$$

If we use the relations

$$(1-T_1)_\alpha = (-T_1)_\alpha + \alpha(1-T_1)_{\alpha-1} \qquad (\alpha > 0)$$

$$(1-T_2)_\beta = (-T_2)_\beta + \beta(1-T_2)_{\beta-1} \qquad (\beta > 0)$$

and remember that (17.4.4.3) is an alternate expression for P we find that

$$P(\tfrac{1}{2}-S_2,\tfrac{1}{2}-S_1,\tfrac{1}{2}-T_2,\tfrac{1}{2}-T_1) - P(\tfrac{1}{2}-S_1,\tfrac{1}{2}-S_2,\tfrac{1}{2}-T_1,\tfrac{1}{2}-T_2)$$

$$= P(S_1,S_2,T_1,T_2) - P(S_2,S_1,T_2,T_1)$$

$$+ \frac{1}{2}\sum_{\alpha+\beta=m-1,\beta>0} \frac{(-1)^{m-1}(2m-1)!}{(2\alpha+1)!(2\beta-1)!}(1-T_1)_\alpha(1-T_2)_{\beta-1}(S_1)_{\alpha+1}(S_2)_\beta$$

$$- \frac{1}{2}\sum_{\alpha+\beta=m-1,\alpha>0} \frac{(-1)^{m-1}(2m-1)!}{(2\alpha-1)!(2\beta+1)!}(1-T_1)_{\alpha-1}(1-T_2)_\beta(S_1)_\alpha(S_2)_{\beta+1} \quad .$$

Changing β into $\beta+1$ in the third term and α into $\alpha+1$ in the fourth term, we see that the two last terms mutually cancel. Hence the lemma.

The proof of (17.4) is now complete.

Case 17.5: We assume now that

$$\rho_1 = \mathrm{Ind}(W, \underline{C}^\times, \chi_1) \ , \ \rho_2 = \mu_2 \oplus \nu_2$$

where χ_1 is a quasi-character of $\underline{C}^\times$. If χ_1 has the form

$$\chi_1(z) = (z\bar{z})^{r_1} z^{m_1} \bar{z}^{n_1} \ ,$$

where m_1 and n_1 are positive or zero integers with only one of them different from zero, and if (μ_1, ν_1) is a pair of quasi-characters of $\underline{R}^\times$ such that

$$\mu_1\nu_1(x) = |x|^{2r_1}(x)^{m_1+n_1} \operatorname{sgn}(x) \ ,$$

$$\mu_1\nu_1^{-1}(x) = (x)^{m_1+n_1} \operatorname{sgn}(x) \ ,$$

we say that the representation $\tau_1 = \mu_1 \oplus \nu_1$ of W is <u>associated</u> to χ_1. By Lemma 15.6 of [1]

(17.5.1) $$\varepsilon'(s, \rho_1, \psi) = \varepsilon'(s, \tau_1, \psi) \ .$$

If η is a quasi-character of $\underline{R}^\times$ and if we substitute to χ_1 the quasi-character $\chi_1 \cdot (\eta \circ N_{\underline{C}/\underline{R}})$ we easily find

(17.5.2) $$\varepsilon'(s, \rho_1 \otimes \eta, \psi) = \varepsilon'(s, \tau_1 \otimes \eta, \psi) \ .$$

In particular,

(17.5.3) $$\varepsilon'(s, \rho_1 \otimes \rho_2, \psi) = \varepsilon'(s, \tau_1 \otimes \rho_2, \psi) \ .$$

Now $\mathfrak{W}(\pi_1, \psi)$ is a subspace of $\mathfrak{W}(\mu_1, \nu_1; \psi)$. Although the latter space is not irreducible under $\mathcal{H}$ the results of Proposition 15.4 are valid for W_i in $\mathfrak{W}(\mu_i, \nu_i; \psi)$, $i = 1,2$. So we see that for W_i in $\mathfrak{W}(\pi_i, \psi)$ the integrals $\Psi(s, W_1, W_2, \Phi)$ and $\tilde{\Psi}(s, W_1, W_2, \Phi)$ are absolutely convergent for $\mathrm{Re}\, s$ large enough, can be analytically continued as meromorphic functions of s and satisfy a functional equation which, by (17.5.3), can be written as

$$\tilde{\Psi}(1-s,W_1,W_2,\hat{\Phi}) = \omega_2(-1)\Psi(s,W_1,W_2,\Phi)\epsilon'(s,\rho_1 \otimes \rho_2,\psi) \quad .$$

So it remains only to show that the assertions 2 and 3 of Theorem 17.2 are true for Ψ with

$$L(s,\pi) = L(s,\rho_1 \otimes \rho_2) \quad .$$

Since there is no harm in replacing the pair (π_1,π_2) by the pair $(\pi_1 \otimes \eta, \pi_2 \otimes \eta^{-1})$ we may assume μ_2 and ν_2 to be of one of the forms

$$\mu_2 = \alpha^{\sigma_2}\xi \quad , \quad \nu_2 = \alpha^{\tau_2} \quad , \tag{17.5.4}$$

$$\mu_2 = \alpha^{\sigma_2} \quad , \quad \nu_2 = \alpha^{\tau_2} \quad . \tag{17.5.5}$$

We treat only the first case, leaving the second one to the reader. We may assume furthermore that $\sigma_2 - \tau_2 \neq 0$. Otherwise if $\sigma = \sigma_2 = \tau_2$ and $\chi_2(z) = (z\bar{z})^{\sigma}$ the representation π_2 is equal to $\pi(\tau_2)$ where τ_2 is the representation $\mathrm{Ind}(W,\underline{C}^{\times},\chi_2)$ and this case will be treated below.

There is a basis W_2^{2m+1}, $m \in \underline{Z}$ of $\mathfrak{W}(\pi_2,\psi)$ satisfying the following conditions (V_- belongs to the Lie algebra, see [1], §5):

$$\pi_2(k(\theta))W_2^{2m+1} = \exp(i(2m+1)\theta)W_2^{2m+1} \quad ,$$

$$\pi_2(V_-)W_2^{2m+1} = W_2^{2m-2} \quad .$$

On the other hand, there is an element W_1 of $\mathfrak{W}(\pi_1,\psi)$ such that

$$\pi_1(k(\theta))W_1 = \exp(i(n_1+1)\theta)W_1 \quad , \quad \pi_1(V_-)W_1 = 0 \quad ,$$

$$\int \varphi_1(a)\,|a|^{s-\frac{1}{2}}d^{\times}a = \int \varphi_1(a)\,|a|^{s-\frac{1}{2}}\mathrm{sgn}(a)d^{\times}a = G_2(s+r_1+n_1) \quad .$$

Clearly we may assume W_1 to be that function and W_2 to be W_2^{2m+1} with $m \geq 0$. Then we may take Φ to be the function

$$\Phi(x,y) = \exp(-\pi(x^2+y^2))(x-iy)^{2m+n_1+2} .$$

Using the duplication formula, we easily find that

$$\Psi(s,W_1,W_2,\Phi)/L(s,\rho_1 \otimes \rho_2)$$

is equal to the difference between the sum

$$\sum_{j,k} \frac{(-4)^{j+k}}{(2j+1)!(2k)!(m-j-k)!}$$

$$\times \quad (s+\tfrac{1}{2}\sigma+\tfrac{1}{2}n_1+j+k+1)_{m-j-k}(s+\sigma_2+r_1+n_1)_{2j+1}(s+\tau_2+r_1+n_1)_{2k} ,$$

and the similar sum obtained by exchanging the roles of σ_2 and τ_2. Here we set

$$\sigma = \sigma_2 + \tau_2 + 2r_1 + n_1 .$$

Hence we obtain, for the above quotient, an expression which is actually a polynomial in s. When $m = 0$, this polynomial reduces to the non-zero constant $\sigma_2 - \tau_2$. So the Theorem 17.2 is completely proved in that case.

Case 17.6: We assume that for $i = 1,2$ the representation π_i has the form $\pi_i = \pi(\rho_i)$ where

$$\rho_i = \mathrm{Ind}(W,\underline{C}^\times,\chi_i)$$

and χ_i is a quasi-character of $\underline{C}^\times$. There is no harm in assuming that χ_i has the form

$$\chi_i(z) = (z\bar{z})^{r_i}\bar{z}^{n_i}$$

where $n_i \geq 0$ is an integer and $n_1 \geq n_2$. Set

$$\mu_i(x) = |x|^{r_i}(x)^{n_i}\,\mathrm{sgn}(x) ,$$

$$\nu_i(x) = (x)^{r_i} ,$$

and $\tau_i = \mu_i \oplus \nu_i$; we see that τ_i is "associated" to χ_i. We claim that

(17.6.1) $$\varepsilon'(s,\rho_1 \otimes \rho_2,\psi) = \varepsilon'(s,\tau_1 \otimes \tau_2,\psi) \quad .$$

Indeed $\rho_1 \otimes \rho_2$ is the representation

$$\mathrm{Ind}(W,\underline{C}^\times,\chi_1\chi_2) \oplus \mathrm{Ind}(W,\underline{C}^\times,\chi_1\chi_2')$$

where χ_2' is the quasi-character

$$\chi_2'(z) = (z\bar{z})^{r_2} z^{n_2} \quad .$$

Moreover, $\xi\mu_1\mu_2 \oplus \nu_1\nu_2$ (respectively $\xi\mu_2\nu_1 \oplus \mu_1\nu_2$) is associated to $\chi_1\chi_2$ (respectively $\chi_1\chi_2'$). Hence the left-hand side of (17.6.1) is equal to

$$\varepsilon'(s,\xi\mu_1\mu_2,\psi)\varepsilon'(s,\xi\nu_1\mu_2,\psi)\varepsilon'(s,\nu_1\nu_2,\psi)\varepsilon'(s,\mu_1\nu_2,\psi) \quad .$$

Since $\mu_1\mu_2 \oplus \nu_1\mu_2$ and $\xi\mu_1\mu_2 \oplus \xi\nu_1\mu_2$ are both associated with the same quasi-character of $\underline{C}^\times$ we may also write this expression in the form

$$\varepsilon'(s,\mu_1\mu_2,\psi)\varepsilon'(s,\nu_1\mu_2,\psi)\varepsilon'(s,\nu_1\nu_2,\psi)\varepsilon'(s,\mu_1\nu_2,\psi)$$

which is the right-hand side of (17.6.1).

Again $\mathfrak{W}(\pi_i,\psi)$ is a subspace of $\mathfrak{W}(\mu_i,\nu_i;\psi)$ and Proposition 17.4 is true for W_i in $\mathfrak{W}(\mu_i,\nu_i;\psi)$. Taking (17.6.1) into account we see that, in the sense of analytical continuation, the functional equation

$$\tilde{\Psi}(1-s,W_1,W_2,\hat{\Phi}) = \omega_2(-1)\varepsilon'(s,\rho_1 \otimes \rho_2,\psi)\Psi(s,W_1,W_2,\Phi)$$

holds for W_i in $\mathfrak{W}(\pi_i,\psi)$.

It remains to see that the assertions 2 and 3 of (17.2) are satisfied with

$$L(s,\pi) = L(s,\rho_1 \otimes \rho_2) \quad .$$

Now $\mathcal{W}(\pi_1,\psi)$ has a basis W_1^n where n is in $\underline{Z}$, $n \geq n_1 + 1$ or $n \leq -n_1 - 1$, and $n \equiv n_1 + 1 \bmod 2$, the elements of which satisfy the following conditions:

$$\pi_1(k(\theta))W_1^n = \exp(in\theta)W_1^n \quad ,$$

$$\pi_1(V_+)W_1^n = W_1^{n+2} \quad \text{for} \quad n \geq n_1 + 1 \; .$$

On the other hand, $\mathcal{W}(\pi_2,\psi)$ contains a vector W_2 such that

$$\pi_2(k(\theta))W_2 = \exp(-i(n_2+1)\theta)W_2 \quad .$$

We may assume that $W_1 = W_1^n$ for some n and W_2 is that function. If φ_i is the function on $\underline{R}^\times$ corresponding to W_i, we know that, if $n \leq -n_1 - 1$,

$$\varphi_1(a) = 0 \quad \text{unless } a > 0 \quad \text{and} \quad \varphi_2(a) = 0 \quad \text{unless} \quad a < 0 \; .$$

It follows that the integral $\Psi(s,W_1,W_2,\Phi)$ vanishes unless $n \geq n_1 + 1$. So we assume that. Then we observe

$$\pi_1(V_+)W_2 = 0 \quad .$$

Using (17.2.5) we see that we may assume $n = n_1 + 1$. Taking Φ to be the function

$$\exp(-\pi(x^2+y^2))(x-iy)^{n_1-n_2}$$

which is permissible, we easily see that $\Psi(s,W_1,W_2,\Phi)$ is a non-zero constant. This completes the proof of (17.2).

§18. Complex case

In this paragraph the ground field is $\underline{C}$. The group K is the group $U(2,\underline{C})$. Then the additive character ψ has the form

$$\psi(x) = \exp(2i\pi(zx+\bar{z}\bar{x})) \text{ , where } z \in \underline{C}^{\times} .$$

We denote by $\mathcal{S}(\underline{C}^2,\psi)$ the subspace of $\mathcal{S}(\underline{C}^2)$ whose elements have the form

$$P(x,\bar{x},y,\bar{y}) \exp(-2\pi (z\bar{z})^{\frac{1}{2}}(x\bar{x}+y\bar{y})) ,$$

where P is a polynomial. It is clear that if Φ belongs to $\mathcal{S}(\underline{C}^2,\psi)$, so does its Fourier transform $\hat{\Phi}$ defined by the formula

$$\hat{\Phi}(x,y) = \int \Phi(u,v)\psi(uy-vx)dudv .$$

Let also π_i , $i = 1,2$, be two irreducible admissible representation of $\mathcal{H}(G,K)$. Let ω_i be the quasi-character of $\underline{C}^{\times}$ such that $\pi_i(a) = \omega_i(a)$ for a in $\underline{C}^{\times}$. We set $\omega = \omega_1\omega_2$. We assume that π_i is infinite dimensional. Then the space $\mathcal{W}(\pi_i,\psi)$ is defined.

For W_i in $\mathcal{W}(\pi_i,\psi)$ and Φ in $\mathcal{S}(\underline{C}^2)$, we define the integrals

$$\Psi(s,W_1,W_2,\Phi) \text{ and } \tilde{\Psi}(s,W_1,W_2,\Phi) .$$

We set $\pi = \pi_1 \times \pi_2$. (This is purely a notational device.)

<u>Theorem 18.1</u>: 1) <u>There is</u> $s_0 \in \underline{R}$ <u>so that for</u> $\mathrm{Re}\, s > s_0$, W_i <u>in</u> $\mathcal{W}(\pi_i,\psi)$ <u>and</u> Φ <u>in</u> $\mathcal{S}(\underline{C}^2)$ <u>the integrals</u> $\Psi(s,W_1,W_2,\Phi)$ <u>and</u> $\tilde{\Psi}(s,W_1,W_2,\Phi)$ <u>are absolutely convergent</u>.

2) <u>There are two Euler factors</u> $L(s,\pi)$ <u>and</u> $L(s,\tilde{\pi})$ <u>with the following properties</u>. <u>For</u> Φ <u>in</u> $\mathcal{S}(\underline{C}^2,\psi)$ <u>set</u>

$$\Psi(s,W_1,W_2,\Phi) = L(s,\pi)\Xi(s,W_1,W_2,\Phi) ,$$

$$\tilde{\Psi}(s,W_1,W_2,\Phi) = L(s,\tilde{\pi})\tilde{\Xi}(s,W_1,W_2,\Phi) .$$

Then $\Xi(s,W_1,W_2,\Phi)$ and $\tilde{\Xi}(s,W_1,W_2,\Phi)$ have the form

$$P(s)(z\bar{z})^{-2s}$$

where P is a polynomial.

3) There are families W_1^j, W_2^j, Φ^j so that

$$\sum_j \Xi(s,W_1^j,W_2^j,\Phi^j) = (z\bar{z})^{-2s} \quad (\text{resp.} \sum_j \tilde{\Xi}(s,W_1^j,W_2^j,\Phi^j) = (z\bar{z})^{-2s}) .$$

4) There is a factor $\epsilon(s,\pi,\psi)$ which is a constant times some power of $(z\bar{z})^{-s}$ such that

$$\tilde{\Xi}(1-s,W_1,W_2,\hat{\Phi}) = \omega_2(-1)\epsilon(s,\pi,\psi)\Xi(s,W_1,W_2,\Phi) .$$

Remarks can be made similar to the ones in §17. We dispense with them.

The Weil group of $\underline{C}$ is just $\underline{C}^\times$. With every finite dimensional, semi-simple representation τ of $\underline{C}^\times$ (i.e., every sum of quasi-characters) we associate with a factor $L(s,\tau)$ and a factor $\epsilon(s,\tau,\psi)$. With every two dimensional representation

$$\tau = \mu \oplus \nu$$

of $\underline{C}^\times$ we associate an irreducible admissible representation $\pi(\tau) = \pi(\mu,\nu)$ of $\mathcal{H}(G_{\underline{C}},K)$.

The rule which gives the factors of Theorem 18.1 is the same as in the real case.

Proposition 18.2: Suppose that $\pi_i = \pi(\tau_i)$ where

$$\tau_i = \mu_i \oplus \nu_i$$

is a two dimensional representation of $\underline{C}^\times$. Then

$$L(s,\tau) = L(s,\tau_1 \otimes \tau_2)$$

$$\epsilon(s,\tau,\psi) = \epsilon(s,\tau_1 \otimes \tau_2,\psi) \quad .$$

The starting point is the following lemma, which is an easy consequence of Barnes' lemma [(17.3.1)].

Lemma 18.2.1: Suppose that for $i = 1,2$ the function φ_i is a continuous function on $\underline{C}^\times$ such that, for Res large enough and all integers A, the integral

$$\int \varphi_i(t)(t\bar{t})^{s-\frac{1}{2}A-\frac{1}{2}}\, t^A\, d^\times t$$

is absolutely convergent. Suppose moreover that this integral vanishes unless $A = A_i$ and takes then the value

$$G_2(s+\sigma_i)G_2(s+\tau_i)$$

where σ_i and τ_i are some constants. Then the integral

$$\int \varphi_1(t)\varphi_2(t)(t\bar{t})^{s-1}\, d^\times t$$

is convergent for Res large enough. It vanishes unless $A_1 + A_2 = 0$ and takes then the value

$$2\pi \frac{G_2(s+\sigma_1+\sigma_2)G_2(s+\sigma_1+\tau_2)G_2(s+\tau_1+\sigma_2)G_2(s+\tau_1+\tau_2)}{G_2(2s+\sigma_1+\tau_1+\sigma_2+\tau_2)} \quad .$$

Here dt is the measure $|dt \wedge d\bar{t}|$ and $d^\times t = dt/(t\bar{t})$.

We state now without proof some simple facts about the representations of $SU(2,\underline{C})$. The irreducible representations of $SU(2,\underline{C})$ are the ρ_n, $n \geq 0$. The space V_n of ρ_n can be regarded as the space of homogeneous polynomials of degree n in two variable X,Y. If

$$k = \begin{pmatrix} \alpha & -\bar{\beta} \\ \beta & \bar{\alpha} \end{pmatrix}$$

is in $SU(2,\underline{C})$ (i.e., if $\alpha\bar{\alpha} + \beta\bar{\beta} = 1$) then

$$\rho_n(k)X^iY^{n-i} = (\alpha X+\beta Y)^i(-\bar{\beta}X+\bar{\alpha}Y)^{n-i}$$

$$= \sum_{0\le t\le n} X^tY^{n-t}\Big[\sum_{j+k=t} \binom{i}{j}\binom{n-i}{k}\alpha^j\,\bar{\alpha}^{n-i-k}\,\beta^{i-j}(-\bar{\beta})^k\Big] .$$

Sometimes it is more convenient to regard V_n as the space of polynomials in X whose degree is at most n. (To obtain the corresponding formulas just substitute 1 to Y).

If $n_1 \ge n_2$ the tensor product $\rho_{n_1} \otimes \rho_{n_2}$ is the direct sum of the ρ_n for $n = n_1 + n_2 - 2k$, $0 \le k \le n_2$. If we regard V_{n_i} as the space of polynomials in X_i whose degree is at most n_i the space $V = V_{n_1} \otimes V_{n_2}$ becomes the space of polynomials in X_1, X_2, whose degree in X_i is at most n_i. If

$$n = n_1 + n_2 - 2k \;, \; 0 \le k \le n_2 \;,$$

up to a constant, there is only one intertwining operator A from V_n to V. It is defined by

(18.3.1) $$\binom{n}{j}AX^{n-j}Y^j = P_j(X_1,X_2)$$

where the P_j's are defined by

(18.3.2) $$(X_1+Z)^{n_1-k}(X_2+Z)^{n_2-k}(X_1-X_2)^k = \sum_{0\le j\le n} Z^jP_j(X_1,X_2) .$$

Explicitely

(18.3.3) $$P_j(X_1,X_2) = (X_1-X_2)^k \sum_{\alpha+\beta=j} X_1^{n_1-k-\alpha}X_2^{n_2-k-\beta}\binom{n_1-k}{\alpha}\binom{n_2-k}{\beta} .$$

On V_n an invariant non-degenerate scalar product is given by

(18.3.4) $$<X^iY^{n-i},X^jY^{n-j}> = 0 \quad \text{if} \quad i+j \neq n$$

$$= (-1)^i/i!j! \quad \text{if} \quad i+j = n \quad .$$

A similar scalar product on $V_{n_1} \otimes V_{n_2}$ is given by

(18.3.5) $$<X_1^{i_1}X_2^{i_2},X_1^{j_1}X_2^{j_2}> = 0 \quad \text{unless } i_1+j_1 = n_1,\ i_2+j_2 = n_2 \ ;$$

$$= (-1)^{i_1+i_2}/i_1!j_1!i_2!j_2! \ , \ \text{if} \quad i_1+j_1 = n_1,$$

$$i_2+j_2 = n_2 \ .$$

We assume now ψ to be of the form

$$\psi(x) = \exp 2i\pi(x+\bar{x}) \quad .$$

Then if Φ is an element of $\mathcal{S}(\underline{C}^2,\psi) \otimes V_n$ such that

$$k.\Phi = \rho_n(k)^{-1}\Phi \ , \ k \in SU(2,\underline{C}) \ ,$$

it must be a linear combination of the functions of the form

$$(x\bar{x}+y\bar{y})^r \exp(-2\pi(x\bar{x}+y\bar{y}))(yX-xY)^m (\bar{x}X+\bar{y}Y)^{n-m} \ , \quad r \geq 0 \ , \ 0 \leq m \leq n \ .$$

Lemma 18.4: Suppose that Φ is the above function. Then let ω be the quasi-character of $\underline{C}^\times$ defined by $\omega(z) = (z\bar{z})^{s-\frac{1}{2}C} z^C$ where $C \in \underline{Z}$. Then the integrals $z(\omega,\Phi)$ and $z(\alpha^2\omega^{-1},\hat{\Phi})$ vanish unless $n = 2m + C$. If this is the case

$$z(\omega,\Phi) = (2\pi)^{-r}(s+\tfrac{1}{2}n)_r \, G_2(s+\tfrac{1}{2}n) \, X^mY^{n-m} \ ,$$

$$z(\alpha^2\omega^{-1},\hat{\Phi}) = (2\pi)^{-r}(s+\tfrac{1}{2}n)_r (i)^C \, G_2(2-2s+\tfrac{1}{2}n) \, X^{n-m}Y^m \quad .$$

Note that $\hat{\Phi}$ transforms under $SU(2,\underline{C})$ according to ρ_n.

Only the last formula needs a proof. First

$$z(\alpha^2\omega^{-1},\hat{\Phi}) = \int \hat{\varphi}(y)(y\bar{y})^2\omega^{-1}(y)d^\times y$$

where $\varphi \in \mathcal{S}(\underline{C})$ is defined by

$$\varphi(x) = \int \Phi(x,v)dv \quad .$$

We find that $\varphi(x)$ is equal to

$$X^{n-m}Y^m e^{-2\pi x\bar{x}} \sum_{k,\theta} (-1)^{k-C}(x\bar{x})^{k+\theta} x^{-C}(2\pi)^{k+\theta-r+m-n}$$

$$\times \binom{m}{n-m-k}\binom{n-m}{k}\binom{r}{\theta}(n-m-k+r-\theta)! \quad .$$

Now

$$z(\alpha^2\omega^{-1},\hat{\Phi}) = i^{|C|}\frac{G_2(2-s+\frac{1}{2}|C|)}{G_2(s-1+\frac{1}{2}|C|)}\int \varphi(x)(x\bar{x})^{s-1-\frac{1}{2}C} x^C d^\times x$$

$$= i^{|C|}\frac{G_2(2-s+\frac{1}{2}|C|)}{G_2(s-1+\frac{1}{2}|C|)} X^{n-m}Y^m \sum_k \binom{m}{n-m-k}\binom{n-m}{k}(-1)^{k-C}(2\pi)^{k+m-n}$$

$$\times \sum_\theta G_2(s-1+k+\theta-\tfrac{1}{2}C)\binom{r}{\theta}(2\pi)^{\theta-r}(n-m-k+r-\theta)! \quad .$$

In this expression the second sum can also be written as:

$$(2\pi)^{-r}G_2(s+k-\tfrac{1}{2}C-1)(n-m-k)!\sum \binom{r}{\theta}(s+k-\tfrac{1}{2}C-1)_\theta \ (n-m-k+1)_{r-\theta}$$

$$= (n-m-k)!\ G_2(s+k-\tfrac{1}{2}C-1)(s+\tfrac{1}{2}n)_r\ (2\pi)^{-r}$$

by the "binomial formula".

So we see that we may assume $r = 0$. If $C \geq 0$ we obtain for $z(\alpha^2\omega^{-1},\hat{\Phi})$ the expression

$$(2\pi)^{-m}X^{n-m}Y^m\, i^C\, G_2(2-s+\tfrac{1}{2}C) \sum_{0\leq h\leq m} (s-1+\tfrac{1}{2}C)_h \binom{m}{h}\frac{(m+C)!}{(h+C)!}(-1)^h \quad .$$

Using (17.1.8) we reduce this to the required expression. The computation is similar for $C \leq 0$.

Let $\pi = \pi(\mu,\nu)$ be an infinite dimensional, irreducible, admissible representation. If

$$\mu(x) = (x\bar{x})^{\sigma-\frac{1}{2}A}x^A, \quad \nu(x) = (x\bar{x})^{\tau-\frac{1}{2}B}x^B, \quad A \in \underline{Z}, \ B \in \underline{Z}, \ A \geq B,$$

the representations ρ_n contained in π are the one for which

$$n = A - B + 2m, \quad m \geq 0 .$$

Let $\mathfrak{W}(\pi,n,\psi)$ be the corresponding subspace. Since the multiplicity of ρ_n in π is one, that space transforms under $SU(2,\underline{C})$ according to ρ_n. In particular, up to a scalar factor, there is one and only one map W from $G_{\underline{C}}$ to V_n such that

$$W(gk) = \rho_n(k)^{-1}W(g) \quad \text{for} \quad k \in SU(2,\underline{C}), \ g \in G_{\underline{C}},$$

and for each v in V_n, the scalar function

$$g \longmapsto \langle W(g), v\rangle$$

belongs to $\mathfrak{W}(\pi,n,\psi)$. We refer to W as the Whittaker function of type ρ_n attached to π. Since

$$W\left[\begin{pmatrix} 1 & x \\ 0 & 1\end{pmatrix} g\right] = \psi(x)\, W(g)$$

the function W is completely defined by the function

$$\varphi: a \longmapsto W\begin{pmatrix} a & 0 \\ 0 & 1\end{pmatrix}$$

from $\underline{C}^\times$ to V_n. In turn φ is completely defined by the following conditions. Let Φ be the element of $\mathcal{S}(\underline{C}^2,\psi) \otimes V_n$ defined by

$$\Phi(x,y) = \exp(-2\pi(x\bar{x}+y\bar{y}))(yX-xY)^m(\bar{x}X+\bar{y}Y)^{A-B+m}$$

and Ψ the element such that

$$\Phi(x,y) = \int \Psi(x,u)\psi(uy)du \quad .$$

Then, for every $s \in \underline{C}$ and p in $\underline{Z}$,

$$\int \varphi(a)(a\bar{a})^{s-\frac{1}{2}(1+p)} a^p d^\times a = \iint \Psi(x,y)(x\bar{x}y\bar{y})^{s-\frac{1}{2}p}(xy)^p \mu(x)\nu(y)\, d^\times x\, d^\times y \ .$$

Using the same technique as in §17 we arrive to the formula

$$\int \varphi(a)(a\bar{a})^{s-\frac{1}{2}(1+p)} a^p d^\times a = (i^{B+p})(2\pi)^{-m}\, m!\,(A-B+m)!\; X^{A+m+p}\, Y^{m-B-p}$$

$$\times \sum_{j,k} \frac{(-2\pi)^{j+k} G_2[s+\sigma+j+\frac{1}{2}(p+A)]\, G_2[s+\tau+k-\frac{1}{2}(p+B)]}{j!\,k!\,(j+A+p)!\,(k-p-B)!\,(m-j-k)!} \ .$$

With our conventions, this means that the integral vanishes unless

$$-A-m \le p \le m - B$$

and is given then by the above sum where j and k are submitted to the following conditions:

$$0 \le j \le m \ , \ 0 \le k \le m \ , \ j+k \le m \ , \ 0 \le A+p+j \ , \ 0 \le k-p-B \ .$$

Now assume that $\pi_i = \pi(\mu_i,\nu_i)$, $i = 1,2$, where

$$\mu_i(x) = (x\bar{x})^{\sigma_i - \frac{1}{2}A_i}\, x^{A_i} \ , \ \nu_i(x) = (x\bar{x})^{\tau_i - \frac{1}{2}B_i}\, x^{B_i} \ , \ A_i - B_i \ge 0 \ .$$

Set

$$n_i = A_i - B_i + 2m_i \ , \quad m_i \ge 0$$

and let W_i be the Whittaker function of type ρ_{n_i} attached to π_i .

Let ρ_n be an irreducible representation of $SU(2,\underline{C})$ contained in the tensor product $\rho_{n_1} \otimes \rho_{n_2}$ and Φ an element of $\mathcal{S}(\underline{C}^2,\psi) \otimes V_n$ such that

$$k.\Phi = \rho_n(k)^{-1}\Phi \quad \text{for} \quad k \in SU(2,\underline{C}) \ .$$

We may consider the integrals

$$(18.5.1)\quad \Psi(s,W_1,W_2,\Phi) = \int_{ZN\backslash G} \langle W_1(g) \otimes W_2(\eta g), z(\alpha^{2s}\omega, g.\Phi)\rangle |\det g|_{\underline{C}}^s\, dg$$

and

(18.5.2) $\tilde{\Psi}(s,W_1,W_2,\Phi)$

$$= \int_{ZN\backslash G} \langle W_1(g) \otimes W_2(\eta g), z(\alpha^{2s}\omega^{-1}, g.\Phi)\rangle |\det g|_{\underline{C}}^{s}\, \omega^{-1}(\det g)\, dg ,$$

where $\langle , \rangle$ is a non-degenerate invariant scalar product on $(V_{n_1} \otimes V_{n_2}) \times V_n$ and

$$\eta = \begin{pmatrix} -1 & 0 \\ 0 & 1 \end{pmatrix} .$$

The Theorem 18.1 makes sense for those new integrals and it amounts to the same to prove it for them. Moreover, if φ_i is the function from $\underline{C}^{\times}$ to V_{n_i} which determines W_i, then:

$$\Psi(s,W_1,W_2,\Phi) = \int_{\underline{C}^{\times}} \langle \varphi_1(a) \otimes \varphi_2(-a), z(\alpha^{2s}\omega,\Phi)\rangle\, |a|_{\underline{C}}^{s-1}\, d^{\times}a ,$$

$$\tilde{\Psi}(s,W_1,W_2,\Phi) = \int_{\underline{C}^{\times}} \langle \varphi_1(a) \otimes \varphi_2(-a), z(\alpha^{2s}\omega^{-1},\Phi)\rangle\, |a|_{\underline{C}}^{s-1}\omega^{-1}(a)\, d^{\times}a .$$

A formal argument shows that we may fix n_2. So we take

$$n_2 = A_2 - B_2 ,\ n_1 = A_1 - B_1 + 2m_1 .$$

If n is such that ρ_n is contained in the tensor product $\rho_{n_1} \otimes \rho_{n_2}$, we can write n in the form:

$$n = n_1 + n_2 - 2r \quad \text{with} \quad 0 \le r \le \operatorname{Inf}(A_2 - B_2, A_1 - B_1 + 2m_1).$$

By Lemma 18.4 we may assume that

$$\Phi(x,y) = \exp(-2\pi(x\bar{x}+y\bar{y}))(yX-xY)^m\, (\bar{x}X+\bar{y}Y)^{n-m} ,\ 0 \le m \le n .$$

Then we find, by the same lemma, that both integrals vanish unless

$$n = 2m + A_1 + A_2 + B_1 + B_2 .$$

So we assume that such is the case. Observe that then

$$m = m_1 - r - B_1 - B_2 \text{ , } n - m = m_1 - r + A_1 + A_2 \text{ .}$$

Using Lemma 18.2.1 we find that

$$(18.5.3) \qquad \int \varphi_1(a)\varphi_2(-a)(a\bar{a})^{s-\frac{1}{2}} d^{\times}a$$

$$= i^{B_1+B_2} c \sum_{j,k}{}' \frac{(-2\pi)^{j+k} \; G_2(S_1'+j)G_2(S_2'+k)}{(m-j-k)!j!k! \; G_2(S_1'+S_2'+j+k)}$$

$$\sum_p (-1)^p X_1^{m_1+A_1+p} X_2^{A_2-p} \frac{1}{(p-B_2)!(A_2-p)!(A_1+p+j)!(k-p-B_1)!}$$

$$G_2(T_2'+j+p+A_1) \; G_2(T_1'+k-p-B_1)$$

where $c > 0$ is some constant and

$$S_1 = \sigma_1 + \sigma_2 + s \text{ , } \quad S_1' = S_1 + \tfrac{1}{2}(A_1+A_2) \text{ ,}$$

$$S_2 = \tau_1 + \tau_2 + s \text{ , } \quad S_2' = S_2 - \tfrac{1}{2}(B_1+B_2) \text{ ,}$$

$$T_1 = \tau_1 + \sigma_2 + s \text{ , } \quad T_1' = T_1 + \tfrac{1}{2}(A_2+B_1) \text{ ,}$$

$$T_2 = \tau_2 + \sigma_1 + s \text{ , } \quad T_2' = T \; - \tfrac{1}{2}(A_1+B_2) \text{ .}$$

We identify $V = V_{n_1} \otimes V_{n_2}$ to the space of all polynomials in X_1 and X_2 of degree at most n_i in X_i.

On the other hand, we may regard Φ as taking its values in the subspace of V which transforms according to ρ_n. We find that

$$z(\alpha^{2s}\omega,\Phi) = \sum_p \xi_p X_1^{m_1-p-B_1} X_2^{p-B_2} G_2(S_1'+S_2'+m_1-r)\binom{n}{n-m}$$

where

$$\sum_p \xi_p X_1^{m_1-p-B_1} X_2^{p-B_2}$$

is the coefficient of $Z^{A_1+A_2+m_1-r}$ in the polynomial $(X_1+Z)^{n_1-r}\ (X_2+Z)^{n_2-r}\ (X_1-X_2)^r$. We thus obtain:

(18.5.6) $\Psi(s,W_1,W_2,\Phi)$

$$= i^{B_1+B_2} c' \sum \frac{(-2\pi)^{j+k}\ G_2(S_1'+j)\ G_2(S_2'+k)\ G_2(S_1'+S_2'+m_1-r)}{j!k!\ (m-j-k)!\ G_2(S_1'+S_2'+j+k)}$$

$$\sum_p (-1)^p \xi_p \frac{(m_1-p-B_1)!(m_1+p+A_1)!}{(k-p-B_1)!(j+p+A_1)!} G_2(T_1'+k-p-B_1)G_2(T_2'+j+A_2) .$$

Here c' is another positive constant.

At this point, we have to distinguish several cases according to the sign of A_1+A_2 , B_1+B_2 , A_1+B_2 , B_1+A_2 . For obvious reasons it is enough to study the following cases:

case number	relative positions of A_i, B_i			
I	$-A_2$	B_1	A_1	$-B_2$
II	$-A_2$	B_1	$-B_2$	A_1
III	$-A_2$	$-B_2$	B_1	A_1

Then we find that

$$L[s,(\mu_1 \oplus \nu_1) \otimes (\mu_2 \oplus \nu_2)] = G_2(S_1')G_2(S_2')G_2(T_1')G_2(T_2')/H(s)$$

where $H(s)$ is given by the following table:

case number	H(s)
I	1
II	$(2\pi)^{-A_1-B_2}\,[T_2-\frac{1}{2}(A_1+B_2)]_{A_1+B_2}$
III	$(2\pi)^{-A_1-B_1-2B_2}[S_2-\frac{1}{2}(B_1+B_2)]_{B_1+B_2}\,[T_2-\frac{1}{2}(A_1+B_2)]_{A_1+B_2}$.

If we denote a priori by $\Xi(s,W_1,W_2,\Phi)$ the quotient of $\Psi(s,W_1,W_2,\Phi)$ by the above L-factor, we find that

$$\Xi(s,W_1,W_2,\Phi) = i^{B_1+B_2}\, c''\, P(S_1,S_2,T_1,T_2)/H(s)$$

where $c'' > 0$ and P is the rational fraction defined in the following lemma:

<u>Lemma 18.6</u>: <u>Let</u> $P(S_1,S_2,T_1,T_2)$ <u>be the rational fraction equal to</u>

$$\sum_{j,k} \frac{(S_1')_j\,(S_2')_k\,(-1)^{j+k}}{j!k!(m_1-j-k)!(S_1'+S_2'+m_1-r)_r}\,(S_1'+S_2'+j+k)_{m_1-j-k}$$

$$\times \quad \sum_p (-1)^p\,\xi_p\,\frac{(A_1+p+m_1)!(m_1-p-B_1)!}{(A_1+p+j)!(k-p-B_1)!}\,(T_1')_{k-p-B_1}\,(T_2')_{j+p+A_1}$$

<u>where</u> S_i <u>and</u> T_i <u>are four variables and</u>

$$S_1' = S_1 + \tfrac{1}{2}(A_1+A_2)\ ,\ S_2' = S_2 - \tfrac{1}{2}(B_1+B_2)\ ,\ T_1' = T_1 + \tfrac{1}{2}(A_2+B_1)$$

$$T_2' = T_2 - \tfrac{1}{2}(A_1+B_2)\ .$$

<u>If</u> $S_1 + S_2 = T_1 + T_2$ <u>then</u> $P(S_1,S_2,T_1,T_2)$ <u>is also equal to</u>

$$\sum_{a+b+c+d=n_1-r} \frac{(-1)^{b+c+A_1}\, r!\,(n_1-r)!\,(n_2-r)!}{a!b!c!d!(m_1-a-b)!(n_1-m_1-c-d)!(m-c-a)!(n-m-b-d)!}$$

$$\times\ [S_1+\tfrac{1}{2}(A_1+A_2)]_a[S_2-\tfrac{1}{2}(B_1+B_2)]_b[1-T_1+\tfrac{1}{2}(A_2+B_1)]_c[1-T_2-\tfrac{1}{2}(A_1+B_2)]_d\ .$$

<u>Moreover, it then satisfies the functional equation</u>

$$P(S_1,S_2,T_1,T_2) = P(1-S_2,1-S_1,1-T_2,1-T_1)\ .$$

Let us show first how the lemma implies Theorem 18.1.

First in the above expression for P it is easily seen that $b \geq B_1 + B_2$, $d \geq A_1 + B_2$. It follows that $\Xi(s,W_1,W_2,\Phi)$ is a polynomial in s. Moreover, it is a non-zero constant if m_1 and r have the value given in the following table:

case number	m_1	r
I	0	A_1-B_1
II	0	$-B_1-B_2$
III	B_1+B_2	0

As for the functional equation, it follows from the lemma and the fact that $\tilde{\Psi}(1-s,W_1,W_2,\hat{\Phi})$ is obtained from $\Psi(s,W_1,W_2,\Phi)$ by changing S_1 into $1-S_2$, S_2 into $1-S_1$, T_1 into $1-T_2$, T_2 into $1-T_1$, A_1 into $-B_1$, B_1 into $-A_1$, A_2 into $-B_2$, B_2 into $-A_2$ and multiplying by

$$i^{A_1+A_2+B_1+B_2}$$

It remains to prove the lemma. We start by writing

$$(S_1'+S_2'+j+k)_{m_1-j-k}(S_1')_j(S_2')_k = \sum_{\lambda+\mu=m_1-j-k} \binom{m_1-j-k}{\lambda} (S_1')_{j+\lambda}(S_2')_{k+\mu} .$$

The expression for P becomes a sum in p and (λ,μ,j,k) where

$$\lambda + \mu + j + k = m_1 .$$

We set

$$\theta = j + \lambda , \ \rho = k + \mu .$$

Then we sum for p , $\theta + \rho = m_1$ and $0 \le j \le \theta$, $0 \le k \le \rho$. We obtain in that way for P the expression

$$\sum_{\theta+\rho=m_1} \frac{(S_1')_\theta \ (S_2')_\rho}{\theta!\rho!(S_1'+S_2'+m_1-r)_r} \sum_p (-1)^p \xi_p \sum_{0\le j\le\theta} (-1)^j \binom{\theta}{j} \Delta_{T_2}^{m_1-j} (T_2')_{m_1+p+A_1}$$

$$\sum_{0\le k\le\rho} (-1)^k \binom{\rho}{k} \Delta_{T_1}^{m_1-k} (T_1')_{m_1-p-B_1} =$$

$$(18.6.1) \quad \sum_{\theta+\rho=m_1} \frac{(S_1')_\theta (S_2')_\rho (-1)^{\theta+\rho}}{\theta!\rho!(S_1'+S_2'+m_1-r)_r} (F_{T_1}F_{T_2})^{m_1} (F_{T_1}^{-1}-1)^\theta (F_{T_2}^{-1}-1)^\rho \, Q(T_1,T_2)$$

where we set

$$Q(T_1,T_2) = \sum_p (-1)^p \xi_p \ (T_1')_{m_1-p-B_1} \ (T_2')_{m_1+A_1+p} .$$

On the other hand, the definition of the ξ_p's amounts to

$$\sum_p (-1)^p \xi_p \ X^{m_1-p-B_1} \ Y^{m_1+A_1+p} = (-1)^{B_2+m} (n_1-r)!(n_2-r)!$$

$$\sum_{\alpha+\gamma=n_1-r} \frac{(-1)^\gamma}{\gamma!(m-\gamma)!\alpha!(n-m-\alpha)!} X^\gamma \ Y^\alpha \ (X+Y)^r .$$

By (17.1.6) we find that

$$Q(T_1,T_2) = (-1)^{B_2+m} (n_1-r)!\ (n_2-r)!$$

$$\sum_{\alpha+\gamma=n_1-r} \frac{(-1)^{\gamma}}{\gamma!(m-\gamma)!\alpha!(n-m-\alpha)!} (T_1')_{\gamma} (T_2')_{\alpha} (T_1'+T_2'+n_1-r)_r \quad .$$

Replacing in (18.6.1) and setting $\theta = a + b$, $\rho = c + d$, we obtain for P the expression

$$q \sum_{\alpha+\gamma=n_1-r,a+b+c+d=m_1} \frac{(S_1')_{a+b} (S_2')_{c+d} (-1)^{b+d+\gamma}}{(S_1' + S_2' + m_1 - r)_r}$$

$$\frac{(T_1'+T_2'+n_1-2m_1-r+a+c)_r\ (T_1'-m_1+a)_{\gamma}\ (T_2'-m_1+c)_{\alpha}}{a!b!c!d!\gamma!(m-\gamma)!\alpha!(n-m-\alpha)!} ,$$

where

$$q = (-1)^{B_1+r} (n_1-r)!\ (n_2-r)! \quad .$$

It is now convenient to sum for

$$a + c = \theta \ , \ b + d = \rho \ , \ \theta + \rho = m_1$$

and to take into account the relation

$$S_1' + S_2' = T_1' + T_2' + n_1 - 2m_1 \quad .$$

The sum in $b + d = \rho$ can be replaced by a close expression with the help of the binomial formula and we obtain by so doing

$$q \sum (-1)^{\gamma+\rho} \frac{(S_1'+S_2'+\theta-r)_r\ (S_1'+S_2'+\theta)_{\rho}}{(S_1' + S_2' + m_1 - r)_r\ \rho!}$$

$$\sum_{a+c=\theta} \frac{(T_1'-m_1+a)_{\gamma}\ (T_2'-m_1+c)_{\alpha}\ (S_1')_a\ (S_1')_c}{a!c!\gamma!\alpha!(m-\gamma)!(n-m-\alpha)!} ,$$

the first summation being for $\alpha + \gamma = n_1 - r$, $\theta + \rho = m_1$.

Now the rational fraction which appears between the two sigmas is in fact equal to

$$(S_1' + S_2' + \theta - r)_{\rho}/\rho! \quad ;$$

which can also be written as

$$\sum_i (-1)^i \binom{r}{i} \frac{(S_1'+S_2'+\theta)_{\rho-i}}{(\rho-i)!} \quad .$$

Replacing and expanding

$$(S_1' + S_2' + \theta)_{\rho-i}$$

with the help of "binomial's formula" we obtainfor P :

$$q \sum_{\alpha+\gamma=n_1-r} (-1)^{\gamma} \sum_{0\le i\le r} \binom{r}{i} \sum_{a+b+c+d=m_1-i} (-1)^{b+d}$$

$$\frac{(T_1'-m_1+a)_{\gamma}\ (T_2'-m_1+c)_{\alpha}\ (S_1')_{a+b}\ (S_2')_{c+d}}{a!b!c!d!\gamma!\alpha!(m-\gamma)!\ (n-m-\alpha)!} \quad .$$

Changing once more variables, we set

$$\theta = a + b \ , \ \rho = c + d \ .$$

Then we obtain

$$q \sum \binom{r}{m_1-\rho-\theta} \frac{(S_1')_{\theta}\ (S_2')_{\rho}}{\theta!\rho!} \sum_{\alpha+\gamma=n_1-r} \frac{(-1)^{\gamma}}{\gamma!\alpha!(m-\gamma)!(n-m-\alpha)!}$$

$$\sum_{a+b=\theta} \frac{(T_1'-m_1+a)_{\gamma}\ (-1)^b\ \theta!}{a!b!} \sum_{c+d=\rho} \frac{(T_2'-m_1+c)_{\alpha}\ (-1)^d\ \rho!}{c!d!} \quad .$$

Now we observe that

$$\sum_{a+b=\theta} \frac{(T_1'-m_1+a)_{\gamma}(-1)^b\theta!}{a!b!} = F_{T_1}^{-\theta}(1 - F_{T_1})^{\theta}\ (T_1' - m_1)_{\gamma}$$

$$= F_{T_1}^{-\theta} (T_1' - m_1)_{\gamma-\theta} \frac{\gamma!}{(\gamma-\theta)!}$$

$$= (-1)^{\gamma+\theta} (1 - T_1' + m_1 - \gamma)_{\gamma-\theta} \frac{\gamma!}{(\gamma-\theta)!} .$$

There is a similar formula for the sum in $c + d = \rho$. All together we obtain for P the following expression:

(18.6.2)
$$\sum \binom{r}{m_1-\rho-\theta} \frac{(S_1')_\theta (S_2')_\rho (-1)^{\theta+\rho}}{\theta!\rho!} R(1 - T_1', 1 - T_2')$$

where R is the polynomial defined by

$$R(T_1,T_2) = (-1)^{A_1} (n_1 - r)! (n_2 - r)!$$

$$\sum_{\alpha+\gamma=n_1-r} \frac{(T_1+m_1-\gamma)_{\gamma-\theta} (T_2+m_1-\alpha)_{\alpha-\rho} (-1)^\gamma}{(m-\gamma)!(n-m-\alpha)!(\gamma-\theta)!(\alpha-\rho)!} .$$

By changing variable this polynomial can also be written as

$$R(T_1,T_2) = (-1)^{m_1+A_1+B_1+A_2} (n_1 - r)!$$

$$\sum_{\alpha+\gamma=n_2-r} \binom{n_2-r}{\alpha} \frac{(T_1+A_2+B_1-\gamma)_{m-\theta-\alpha} (T_2-A_1-B_2-\alpha)_{n-m-\gamma-\rho} (-1)^\gamma}{(m-\theta-\alpha)! (n-m-\gamma-\rho)!}$$

$$= (-1)^{m_1+A_1+B_1+A_2} (F_{T_2} \Delta_{T_1} - F_{T_1} \Delta_{T_2})^{n_2-r}$$

$$\frac{(T_1+A_2+B_1)_{m-\theta} (T_2-A_1-B_2)_{n-m-\rho}}{(m-\theta)! (n-m-\rho)!} .$$

But

$$F_{T_2} \Delta_{T_1} - F_{T_1} \Delta_{T_2} = \Delta_{T_1} - \Delta_{T_2} .$$

Therefore an alternate formula for R is

$$R(T_1,T_2) = (-1)^{m_1+A_1+B_1+A_2} (n_1 - r)!$$

$$\sum_{\alpha+\gamma=n_2-r} \binom{n_2-r}{\alpha} \frac{(T_1+A_2+B_1)_{m-\theta-\alpha}(T_2-A_1-B_2)_{n-m-\rho-\gamma}}{(m-\theta-\alpha)!\,(n-m-\rho-\gamma)!} (-1)^{\gamma} .$$

Now we replace R by this expression in (18.6.2) and set

$$a = \theta \ , \ b = \rho \ , \ c = m - \theta - \alpha \ , \ d = n - m - \rho - \gamma \ .$$

We obtain a sum for

$$a + b + c + d = n_1 - r$$

which is just the expression given for P in Lemma 18.6.

To prove the functional equation given in the lemma, we merely observe that, when $S_1 + S_2 = T_1 + T_2$ the initial expression for P can also be written in the form

$$(-1)^{A_1-B_1} \sum_{j,k} \frac{(T_1')_j\,(T_2')_k\,(-1)^{j+k}}{j!k!(m_1'-j-k)!(T_1'+T_2'+m_1'-r)_r} (T_1'+T_2'+j+k)_{m_1'-j-k}$$

$$\sum_p (-1)^p \xi_p \frac{(m_1'-p-A_1)!\,(m_1'+B_1+p)!}{(k-p-A_1)!\,\,(j+B_1+p)!} (S_1')_{k-p-A_1} (S_2')_{j+p+B_1} ,$$

where

$$m_1' = m_1 + A_1 - B_1 \ .$$

This can be transformed, following the above method, into a new expression on which the functional equation is obvious.

This concludes the proof of Lemma 18.6 and Theorem 18.1. Phew!

Chapter V: Global Theory for GL(2) x GL(2)

§19. Global functional equation for GL(2) x GL(2)

In this chapter, the ground field F is an A-field. As usual we denote by $\underline{A}$ its ring of adeles and $\underline{I}$ its group of ideles. The set Ω of all quasi-characters of $\underline{I}/F^{\times}$ has a natural structure of one dimensional complex analytic variety (with infinitely many connected components). So we may speak of homomorphic or meromorphic functions defined on Ω (or an open subset) with values in $\underline{C}$ or, more generally, in a complete locally convex space. However, rather than speak of a meromorphic map from Ω to a space of distributions or functions, we shall speak of a meromorphic family of distributions or functions depending on a quasi-character.

For instance, there is a family

$$\Phi \longrightarrow z(\omega,\Phi)$$

of distributions on $\mathcal{S}(\underline{A}^2)$, meromorphic with respect to ω, which if $|\omega| = \alpha_F^s$ and $s > 1$ is defined by the convergent integral

$$z(\omega,\Phi) = \int_{\underline{I}} \Phi(0,t)\omega(t)d^{\times}t .$$

If Φ is in $\mathcal{S}(\underline{A}^2)$ we define a meromorphic family of functions on $G_{\underline{A}}$, depending on (μ_1,μ_2) in $\Omega \times \Omega$ by the formula

$$f(g) = z(\alpha\mu_1\mu_2^{-1},g.\Phi)\ \mu_1(\det g)\,|\det g|^{\frac{1}{2}} . \tag{19.1}$$

Then f satisfies the relation

$$f\left[\begin{pmatrix} a & x \\ 0 & b \end{pmatrix} g\right] = |a/b|^{\frac{1}{2}}\ \mu_1(a)\mu_2(b)f(g) .$$

In particular, if P is the group of triangular matrices, then

$$f(\gamma g) = f(g) \quad \text{for } \gamma \text{ in } P_F .$$

For a given Φ in $\mathcal{S}(\underline{A}^2)$ we consider the series

$$E(\Phi,\mu_1,\mu_2) = \sum_{P_F\backslash G_F} f(\gamma) \tag{19.2}$$

where f is the function 19.1. We have the following result.

<u>Proposition 19.3</u>: <u>Suppose that</u> $|\mu_1\mu_2^{-1}| = \alpha^s$ <u>with</u> $s > 1$. <u>Then the series 19.2 is absolutely convergent and defines a distribution on</u> $\mathcal{S}(\underline{A}^2)$. <u>It extends into a meromorphic family of distributions, depending on</u> (μ_1,μ_2) in $\Omega \times \Omega$. <u>As such it satisfies the functional equation</u>

$$E(\Phi,\mu_1,\mu_2) = E(\hat{\Phi},\mu_2,\mu_1) .$$

Here we identify $V = F^2$ to its dual by

$$<(x,y),(u,v)> = yu - xv .$$

Accordingly, we define the Fourier transform of a Φ in $\mathcal{S}(\underline{A}^2)$ by

$$\hat{\Phi}(x,y) = \int \Phi(u,v)\psi(yu-xv)dudv .$$

Of course, ψ is a nontrivial character of $\underline{A}/F$ and du or dv is the self dual Haar measure on $\underline{A}$. In particular,

$$(\hat{\Phi})^{\wedge} = \Phi .$$

Take $|\mu_1\mu_2^{-1}| = \alpha^s$ with $s > 1$. Then f is defined by a convergent integral. Computing formally at first and replacing f by its value in 19.2 we find

$$E(\Phi,\mu_1,\mu_2) = \sum_{\xi\in F} f\left[w\begin{pmatrix}1 & \xi\\ 0 & 1\end{pmatrix}\right] + f(e)$$

$$= \sum_{\xi\in F}\int_{\underline{I}} \Phi(-t,-t\xi)\,|t|\mu_1\mu_2^{-1}(t)d^{\times}t + \int_{\underline{I}} \Phi(0,t)\,|t|\mu_1\mu_2^{-1}(t)d^{\times}t.$$

Replacing the integration on $\underline{I}$ by a summation on $F^{\times}$ followed by an integration on $\underline{I}/F^{\times}$ and exchanging the order of summation and integration we obtain

$$\int_{\underline{I}/F^{\times}} \sum_{(\xi,\eta)\neq(0,0)} \Phi[a(\xi,\eta)]\,|a|\,\mu_1\mu_2^{-1}(a)d^{\times}a \quad .$$

We use the notions introduced in [8] §11 except that we write $\theta^0(\alpha^s\mu,\Phi)$, $\theta^1(\alpha^s\mu,\Phi)$, $\lambda(\alpha^s\mu)$ for $\theta^0(\mu,s,\Phi)$, $\theta^1(\mu,s,\Phi)$, $\lambda(s,\mu)$ respectively. (See also Summary and Notations). Hence

$$E(\Phi,\mu_1,\mu_2) = \theta^0(\alpha\mu_1\mu_2^{-1},\Phi) + \theta^1(\alpha\mu_1\mu_2^{-1},\Phi) \quad .$$

We know that the right-hand side is defined (by absolutely convergent double integrals) if, as we assume, $s > 1$. Therefore, it follows that the formal computations are justified and that the series (19.2) is absolutely convergent for $\mathrm{Re}\, s > 1$. Also, by Poisson formula,

$$(19.4) \qquad E(\Phi,\mu_1,\mu_2) = \theta^0(\alpha\mu_1\mu_2^{-1},\Phi) + \theta^0(\alpha\mu_2\mu_1^{-1},\hat{\Phi}) - \lambda(\alpha\mu_2\mu_1^{-1})\hat{\Phi}(0) - \lambda(\alpha\mu_1\mu_2^{-1})\Phi(0) \quad .$$

This gives the analytic continuation and the functional equation.

If we substitute to Φ the function $g.\Phi$ we find the "Eisenstein series":

$$E(g.\Phi,\mu_1,\mu_2)\mu_1(\det g)\,|\det g|^{\frac{1}{2}} = \sum_{\gamma} f(\gamma g) \quad .$$

Since the Fourier transform of the function $g.\Phi$ is the function

$$|\det g|^{-1}\, g'.\hat{\Phi} \quad \text{where} \quad g' = \begin{pmatrix} \det g^{-1} & 0 \\ 0 & \det g^{-1} \end{pmatrix} g \quad ,$$

we easily find the functional equation

(19.5) $\quad E(g.\Phi,\mu_1,\mu_2)\mu_1(\det g)|\det g|^{\frac{1}{2}} = E(g.\hat{\Phi},\mu_2,\mu_1)\mu_2(\det g)|\det g|^{\frac{1}{2}}$.

Also we have

(19.6) $\quad E(g.\Phi,\mu_1,\mu_2)\mu_1(\det g)|\det g|^{\frac{1}{2}} = \theta^0(\alpha\mu_1\mu_2^{-1},g.\Phi)\mu_1(\det g)|\det g|^{\frac{1}{2}}$

$$+ \theta^0(\alpha\mu_2\mu_1^{-1},g'.\hat{\Phi})\mu_1(\det g)|\det g|^{\frac{1}{2}}$$

$$- \lambda(\alpha\mu_2\mu_1^{-1})\hat{\Phi}(0)\mu_1(\det g)|\det g|^{\frac{1}{2}}$$

$$- \lambda(\alpha\mu_1\mu_2^{-1})\Phi(0)\mu_1(\det g)|\det g|^{\frac{1}{2}} .$$

Now let ω be a quasi-character of $\underline{I}/F^{\times}$ and φ a continuous function on $G_{\underline{A}}$ which satisfies the following conditions:

$\varphi(a\gamma g) = \omega(a)\varphi(g)$ for a in $\underline{I}$, γ in G_F,

φ is K-finite on the right, K being the standard maximal compact subgroup,

φ is a rapidly decreasing function, if F is a number field and compactly supported modulo $Z_{\underline{A}}G_F$, if F is a function field.

We consider the following integrals:

(19.7) $\quad \int_{G_F Z_{\underline{A}}\backslash G_{\underline{A}}} \varphi(g)E(g.\Phi,\mu_1,\mu_2)\mu_1(\det g)|\det g|^{\frac{1}{2}}\,dg$,

(19.8) $\quad \int_{G_F Z_{\underline{A}}\backslash G_{\underline{A}}} \varphi(g)E(g.\hat{\Phi},\mu_2,\mu_1)\mu_2(\det g)|\det g|^{\frac{1}{2}}\,dg$,

where

$$\mu_1 = \alpha^{s-\frac{1}{2}} , \quad \mu_2 = \alpha^{\frac{1}{2}-s}\,\omega^{-1} .$$

Note that the integrands are invariant on the left under $G_F Z_{\underline{A}}$.

Suppose that s is so chosen that

$$\alpha^{2-2s} \neq \omega \quad \text{and} \quad \alpha^{2s} \neq \omega^{-1} .$$

Then the Eisenstein series has no poles. In (19.7) we may substitute (19.6) for it and integrate on $G_F \backslash G'$ (where G' has the same meaning as in the Summary - Notations.) We find that (19.7) is absolutely convergent. In general, it defines a meromorphic function of s (or α^s). The poles (if any) are simple and occur for

$$\alpha^{2-2s} = \omega \quad \text{and} \quad \alpha^{2s} = \omega^{-1} .$$

At $\alpha^{2-2s} = \omega$ the residue is proportional to

$$\hat{\Phi}(0) \int_{G_F Z_{\underline{A}} \backslash G_{\underline{A}}} |\det g|^{s-1} \varphi(g) dg .$$

Of course, if this quantity vanishes, the function is actually holomorphic at the point in question. At $\alpha^{2s} = \omega^{-1}$, the residue is proportional to

$$\Phi(0) \int_{G_F Z_{\underline{A}} \backslash G_{\underline{A}}} |\det g|^{s} \varphi(g) dg .$$

Finally, in the number field case, we find also that (19.7) is bounded at infinity in vertical strips of finite width.

Similar considerations apply to the integral (19.8). By (19.5) both integrals are equal.

To go further, we take Res sufficiently large and replace in (19.7) the "Eisenstein series" by its expression as a series. We obtain that (19.7) is equal to

$$\int_{G_F Z_{\underline{A}} \backslash G_{\underline{A}}} \varphi(g) \sum_{P_F \backslash G_F} f(\gamma g) dg = \int_{P_F Z_{\underline{A}} \backslash G_{\underline{A}}} \varphi(g) f(g) dg \quad .$$

We may set

$$g = \begin{pmatrix} 1 & x \\ 0 & 1 \end{pmatrix} \begin{pmatrix} a & 0 \\ 0 & 1 \end{pmatrix} k$$

where x varies in $\underline{A}/F$, a in $\underline{I}/F^{\times}$ and k in K . Then, in the number field case, for each N , there is a constant c such that

$$|\varphi(g)| \leq c |a|^{-N} \quad .$$

In the function field case, there is a constant c so that

$$\varphi(g) = 0 \quad \text{if} \quad |a| > c \quad .$$

Moreover we get

$$dg = dx \, |a|^{-1} \, d^{\times}a \, dk$$

and

$$f(g) = |a|^{s} \, z(\alpha^{2s}\omega, k.\Phi) \quad .$$

So (19.7) is equal to the new integral

$$\int_{F\backslash \underline{A} \times F^{\times} \backslash \underline{I} \times K} \varphi\left[\begin{pmatrix} 1 & x \\ 0 & 1 \end{pmatrix}\begin{pmatrix} a & 0 \\ 0 & 1 \end{pmatrix} k\right] z(\alpha^{2s}\omega, k.\Phi) \, |a|^{s-1} \, d^{\times}a \, dx dk \quad .$$

Obviously, this integral is convergent if $\operatorname{Re} s$ is sufficiently large. So the above computation is then justified. In the above integral, we may integrate first on $F\backslash\underline{A}$. We obtain then the new integral

$$\Lambda(s,\varphi,\Phi) = \int_{F^{\times}\backslash \underline{I} \times K} \varphi^{0}\left[\begin{pmatrix} a & 0 \\ 0 & 1 \end{pmatrix} k\right] z(\alpha^{2s}\omega, k.\Phi) \, |a|^{s-1} \, d^{\times}a \, dk$$

where φ^{0} is the constant term of φ , i.e.,

$$\varphi^0(g) = \int_{\underline{A}/F} \varphi\left[\begin{pmatrix}1 & x\\ 0 & 1\end{pmatrix} g\right] dx \ .$$

Similar considerations apply to the integral

(19.10)

$$\tilde{\Lambda}(s,\varphi,\Phi) = \int_{F^\times\backslash \underline{I}\times K} \varphi^0\left[\begin{pmatrix}a & 0\\ 0 & 1\end{pmatrix} k\right] z(\alpha^{2s}\omega^{-1}, k.\Phi)\ |a|^{s-1}\omega^{-1}(a)d^\times a\ dk \ .$$

In particular, if Res is small enough, the integral $\tilde{\Lambda}(1-s,\varphi,\hat{\Phi})$ is convergent and equal to (19.8).

Finally, we have obtained the following result.

Proposition 19.11: If Res is large enough the integrals $\Lambda(s,\varphi,\Phi)$ and $\tilde{\Lambda}(s,\varphi,\Phi)$ are absolutely convergent. They can be analytically continued as meromorphic functions of s in the whole complex plane. As such they satisfy the functional equation:

$$\Lambda(s,\varphi,\Phi) = \tilde{\Lambda}(1-s,\varphi,\hat{\Phi}) \ .$$

Moreover, if F is a number field, and Φ is fixed, the meromorphic function $\Lambda(s,\varphi,\Phi)$ is bounded at infinity in vertical strips of finite width.

Alternatively we may say that we have meromorphic families of distributions. The poles of the family $\Lambda(s,\varphi,\Phi)$ occur for

$$\alpha^{2-2s} = \omega \quad \text{and} \quad \alpha^{2s} = \omega^{-1} \ .$$

Consider now the following situation. For $i = 1,2$ we let ω_i be a quasi-character of $\underline{I}/F^\times$ and V_i an invariant irreducible subspace of $\mathcal{C}_0(\omega_i)$. We call π_i the class of the representation of the Hecke algebra $\mathcal{H}$ on V_i. We take φ_i in V_i and apply the previous results to

$$\varphi = \varphi_1\varphi_2 \ , \ \omega = \omega_1\omega_2 \ .$$

Since a cusp form is rapidly decreasing if F is a number field, and compactly supported if F is a function field, the function φ satisfied the above assumptions. Moreover, we know that

$$\varphi_i(g) = \sum_{\alpha \in F^\times} W_i\left[\begin{pmatrix} \alpha & 0 \\ 0 & 1 \end{pmatrix} g\right]$$

where W_i belongs to $\mathfrak{W}(\pi_i,\psi)$. Then

$$\varphi^0(g) = \sum_{\alpha \in F^\times} W_1\left[\begin{pmatrix} \alpha & 0 \\ 0 & 1 \end{pmatrix} g\right] W_2\left[\begin{pmatrix} -\alpha & 0 \\ 0 & 1 \end{pmatrix} g\right] .$$

If we substitute this into (19.9) and replace the summation on $F^\times$ and the integration on $\underline{I}/F^\times$ by an integration on $\underline{I}$ we obtain that

$$\Lambda(s,\varphi,\Phi) = \Psi(s,W_1,W_2,\Phi)$$

where we set

$$\Psi(s,W_1,W_2,\Phi) = \int_{\underline{I}\times K} W_1\left[\begin{pmatrix} a & 0 \\ 0 & 1 \end{pmatrix} k\right] W_2\left[\begin{pmatrix} -a & 0 \\ 0 & 1 \end{pmatrix} k\right] z(\alpha^{2s}\omega,k.\Phi)\,|a|^{s-1}d^\times a\,dk \ .$$

Note that this integral can also be thought of as an integral on the quotient space $N_{\underline{A}}Z_{\underline{A}}\backslash G_{\underline{A}}$:

$$\Psi(s,W_1,W_2,\Phi) = \int_{N_{\underline{A}}Z_{\underline{A}}\backslash G_{\underline{A}}} W_1(g)W_2(\eta g)\ z(\alpha^{2s}\omega,g.\Phi)\,|\det g|^s\,dg \ .$$

The following lemma shows that if $\mathrm{Re}\,s$ is large enough the above computation is justified.

<u>Lemma 19.12</u>: <u>If</u> $\mathrm{Re}\,s$ <u>is large enough the integral</u> $\Psi(s,W_1,W_2,\Phi)$ <u>is</u>

absolutely convergent.

There is no harm in assuming that

$$W_i(g) = \prod W_{iv}(g_v) ,$$

$$\Phi(x,y) = \prod \Phi_v(x_v,y_v) ,$$

where, for all v, the function W_{iv} belongs to $\mathfrak{W}(\pi_{iv},\psi_v)$ and the function Φ_v to $\mathcal{S}(F_v^2)$. Of course, for almost all v the representation π_{iv} contains the unit representation of K_v, the function W_{iv} is invariant under K_v and takes the value one on K_v and Φ_v is the characteristic function of R_v^2. On the other hand, the space $N_{\underline{A}}Z_{\underline{A}}\backslash G_{\underline{A}}$ is the restricted product of the spaces $N_vZ_v\backslash G_v$. We may assume that the measure on the adelic space is the product of the local invariant measures, these, for v nonarchimedean, being chosen as in Proposition 15.9. Using this proposition and standard arguments, the lemma is easily obtained. (Cf. [1], p. 356). Moreover, it is found that for $\operatorname{Re} s$ large enough the integrals

$$\Psi(s,W_{1v},W_{2v},\Phi_v)$$

and the infinite product

$$\prod_v \Psi(s,W_{1v},W_{2v},\Phi_v)$$

are absolutely convergent, the infinite product being equal to $\Psi(s,W_1,W_2,\Phi)$.

Similar results hold for the integral

$$\tilde{\Psi}(s,W_1,W_2,\Phi) = \int_{\underline{I}\times K} W_1\left[\begin{pmatrix} a & 0 \\ 0 & 1\end{pmatrix}k\right]W_2\left[\begin{pmatrix} -a & 0 \\ 0 & 1\end{pmatrix}k\right]z(\alpha^{2s}\omega^{-1},k.\Phi)\,|a|^{s-1}\omega^{-1}(a)\,d^\times a\,dk$$

$$= \int_{N_{\underline{A}}Z_{\underline{A}}\backslash G_{\underline{A}}} W_1(g)W_2(\eta g)\,z(\alpha^{2s}\omega^{-1},g.\Phi)\,|\det g|^s\omega^{-1}(\det g)\,dg .$$

Now we may restate the results of 19.11 in the following form:

Theorem 19.13: For $i = 1,2$ let π_i be an irreducible component of $\mathcal{A}_0(\omega_i)$. For W_i in $W(\pi_i,\psi)$ and Φ in $\mathcal{S}(\underline{A}^2)$ the integrals $\Psi(s,W_1,W_2,\Phi)$ and $\tilde{\Psi}(s,W_1,W_2,\Phi)$ are absolutely convergent if $\operatorname{Re} s$ is large enough. The integrals can be analytically continued and define meromorphic families of distributions on $\mathcal{S}(\underline{A}^2)$. As such they satisfy the functional equation

$$\Psi(s,W_1,W_2,\Phi) = \tilde{\Psi}(1-s,W_1,W_2,\hat{\Phi}) .$$

Now let π be the representation $\pi_1 \times \pi_2$ (as in the archimedean case, we use this just as a notational device). Set

$$L(s,\pi) = \prod_v L(s,\pi_v) = \prod_v L(s,\pi_{1v} \times \pi_{2v}) .$$

It follows from Proposition (15.9) (or Theorem 19.13) that for $\operatorname{Re} s$ large enough all the factors are holomorphic functions of s and their product is absolutely convergent. Hence $L(s,\pi)$ is holomorphic in some (right) half space. The same is true of

$$L(s,\tilde{\pi}) = \prod_v L(s,\tilde{\pi}_v) = \prod_v L(s,\tilde{\pi}_{1v} \times \tilde{\pi}_{2v}) .$$

In the infinite product

$$\epsilon(s,\pi) = \prod_v \epsilon(s,\pi_v,\psi_v) = \prod_v \epsilon(s,\pi_{1v} \times \pi_{2v},\psi_v)$$

almost all factors are equal to one. Moreover it follows from 14.8.5 that the product is actually independent of ψ. As a function of s the factor $\epsilon(s,\pi)$ is just an exponential times a constant.

Following step by step the proof of 11.1.3 in [1], we arrive to the following theorem.

<u>Theorem 19.14</u>: <u>Under the above assumptions, the Euler products</u> $L(s,\pi)$ <u>and</u> $L(s,\tilde{\pi})$ <u>are absolutely convergent for</u> $\mathrm{Re}\,s$ <u>large enough</u>. <u>They can be analytically continued as meromorphic functions of</u> s <u>in the whole complex plane</u>. <u>As such they satisfy the functional equation</u>

$$L(s,\pi) = \epsilon(s,\pi)L(1-s,\tilde{\pi}) \ .$$

Actually, we find that the functional equation is

$$L(s,\pi) = \omega_2(-1)\epsilon(s,\pi)L(1-s,\tilde{\pi}) \ .$$

But $\omega_2(-1)$ is one.

We have also some information on the poles. They are simple and can occur only for

$$\alpha^{2-2s} = \omega \quad \text{and} \quad \alpha^{2s} = \omega^{-1} \ .$$

So if ω is not principal (of the form α_F^{σ}) the function $L(s,\pi)$ is actually entire.

In general, if F is a number field, the function $L(s,\pi)$ is bounded at infinity in vertical strips of finite width. If F is a function field, it is a rational function of Q^{-s} (where $Q > 1$ is a generator of the subgroup of $\underline{R}_+^\times$ generated by the $|a|$). This follows actually from the method which gives the analytic continuation. In the function field case, one can also use the method given in [7] Theorem 4,VIII, p.130.

An application of these results is the following. Suppose that K is a separable quadratic extension of F. Let χ be a quasi-character of $K_{\underline{A}}^\times/K^\times \simeq W_{K/K}$ and (cf. [1] §12)

$$\sigma = \mathrm{Ind}(W_{K/F}, W_{K/K}, \chi) \ .$$

Then, with the notations of [1], §12, for each place v of F, σ_v is a representation of W_{F_v} and the representation $\pi(\sigma_v)$ is defined. Call $\pi(\sigma)$ or $\pi(\chi)$ the representation $\otimes_v \pi(\sigma_v)$.

Corollary 19.15: Suppose that ω is a quasi-character of $F_A^\times/F^\times$ and π an irreducible component of $\mathcal{A}_0(\omega)$. Let K be a separable quadratic extension of F . For each quasi-character χ of $K_A^\times/K^\times$ the infinite Euler product

$$L(s,\pi \times \pi(\chi))$$

is holomorphic in some right half space, can be analytically continued as a meromorphic function of s and satisfy the functional equation

$$L(s,\pi \times \pi(\chi)) = \epsilon(s,\pi \times \pi(\chi))\, L(1-s,\tilde{\pi} \times \pi(\chi^{-1})) .$$

Let ζ be the quasi-character of $F_A^\times/F^\times$ attached to K . Then if the product

$$(\omega\zeta) \cdot (\chi\,|F_A^\times)$$

is not a principal character, $L(s,\pi \times \pi(\chi))$ is an entire function of s , bounded in vertical strips (if F is a number field).

If χ is not of the form

$$\chi = \mu \circ N_{K/F}$$

the representation $\pi(\chi)$ is contained in the space of cusp forms (for the group $GL(2,F_A)$); we may therefore apply Theorem 19.14 with π replacing π_1 , $\pi(\chi)$ replacing π_2 , ω replacing ω_1 and

$$(\omega\,\zeta) \cdot (\chi\,|F_A^\times)$$

replacing ω_2 .

If χ is of the form

$$\chi = \mu \circ N_{K/F}$$

where μ is a quasi-character of $F_{\underline{A}}^{\times}/F^{\times}$ then

$$\sigma = \mu \oplus \mu\zeta ,$$

and (cf. 15.1, 17.3, 18.2)

$$L(s,\pi \times \pi(\chi)) = L(s,\pi \otimes \mu)L(s,\pi \otimes \mu\zeta) .$$

So all we have to do is to apply the results of [1], Corollary 11.2.

(19.15.1): Suppose that v is a nonarchimedean place of F which does not split in K. Let w be the unique place of v lying above v, then ζ_v is the character of $F_v^{\times}$ attached to the quadratic extension K_w. If

$$\omega_v\zeta_v \cdot (\chi_w|F_v^{\times})$$

is ramified, the supplementary condition of 19.15 is surely satisfied $L(s,\pi \times \pi(\chi))$ is an entire function.

Another application is the following.

Corollary 19.16: Suppose that σ and τ are two irreducible representations of the Weil group W_K of degree two. Suppose that for each place v of F the representations $\pi(\sigma_v)$ and $\pi(\tau_v)$ are defined, and that the representations

$$\pi(\sigma) = \underset{v}{\otimes}\, \pi(\sigma_v) , \quad \pi(\tau) = \underset{v}{\otimes}\, \pi(\tau_v)$$

occur in the space of parabolic forms. Then, for all places v of F, the following equality hold:

$$L(1-s,\tilde{\pi}(\sigma_v) \times \tilde{\pi}(\tau_v))\varepsilon(s,\pi(\sigma_v) \times \pi(\tau_v),\psi_v)/L(s,\pi(\sigma_v) \times \pi(\tau_v))$$

$$= L(1-s,\tilde{\sigma}_v \otimes \tilde{\tau}_v)\varepsilon(s,\sigma_v \otimes \tau_v,\psi_v)/L(s,\sigma_v \otimes \tau_v) .$$

The left-hand side of this équality is what we denote by

$$\epsilon'(s,\pi(\sigma_v) \times \pi(\tau_v),\psi_v) .$$

Similarly, it is convenient to denote the right-hand side by

$$\epsilon'(s,\sigma_v \otimes \tau_v,\psi_v) .$$

Consider the equalities

$$(19.16.1) \qquad L(s,\pi(\sigma_v) \times \pi(\tau_v)) = L(s,\sigma_v \otimes \tau_v) ,$$

$$L(s,\tilde{\pi}(\sigma_v) \times \tilde{\pi}(\tau_v)) = L(s,\tilde{\sigma}_v \otimes \tilde{\tau}_v) ,$$

$$\epsilon(s,\pi(\sigma_v) \times \pi(\tau_v),\psi_v) = \epsilon(s,\sigma_v \otimes \tau_v,\psi_v) .$$

By Theorems 17.3 and 18.2, they are surely true for all archimedean places. If v is a nonarchimedean place and

$$\tau_v = \mu \oplus \nu$$

where μ and ν are quasi-characters of $F_v^{\times}$ (or the local Weil group) they are also true by Theorem 15.1. Since this is the case for almost all v , we see that there is a finite set S of nonarchimedean places such that, for v not in S , the relations (19.6.1) hold. In particular, for all v not in S ,

$$(19.16.2) \qquad \epsilon'(s,\pi(\sigma_v)\times \pi(\tau_v),\psi_v) = \epsilon'(s,\sigma_v \otimes \tau_v,\psi_v) .$$

Now if we compare the functional equations

$$L(1-s,\tilde{\sigma} \otimes \tilde{\tau}) = \epsilon(s,\sigma \otimes \tau)L(s,\sigma \otimes \tau)$$

and

$$L(1-s,\tilde{\pi}(\sigma) \times \tilde{\pi}(\tau)) = \epsilon(s,\pi(\sigma) \times \pi(\tau))L(s,\pi(\sigma) \times \pi(\tau))$$

and take into account the relations (19.16.1) we arrive at

$$(19.16.3) \qquad \prod_{v\in S} \epsilon'(s,\pi(\sigma_v) \times \pi(\tau_v),\psi_v) = \prod_{v\in S} \epsilon'(s,\sigma_v \otimes \tau_v,\psi_v) .$$

So the corollary is proved if S has at most one element. If not, fix a place w in S. There are quasi-characters χ of $\underline{I}/F^{\times}$ such that $\chi_w = 1$ and the order m_v of χ_v for v in S, $v \neq w$, is arbitrarily large. ([1], Lemma 12.5, p.404). If m_v is sufficiently large

$$(19.16.4)\qquad \epsilon'(s,\sigma_v \otimes (\tau_v \otimes \chi_v),\psi_v) = \epsilon'(s,\chi_v \det\tau_v \det\sigma_v,\psi_v)[\epsilon'(s,\chi_v,\psi_v)]^3,$$

$$v \in S\ ,\ v \neq w\ .$$

(Cf. [9] appendix). On the other hand, for a in $F_v^{\times}$,

$$\pi(\sigma_v)(a) = \det\sigma_v(a)\ ,\ \pi(\tau_v)(a) = \det\tau_v(a)\ .$$

Therefore by Theorem 16.1 we find that if m_v is sufficiently large

(19.16.5)

$$\begin{aligned}\epsilon'(s,\pi(\sigma_v) \times \pi(\tau_v \otimes \chi_v),\psi_v) &= \epsilon'(s,\pi(\sigma_v) \times [\pi(\tau_v) \otimes \chi_v],\psi_v)\\ &= \epsilon'(s,\chi_v \det\tau_v \det\sigma_v,\psi_v)\epsilon'(s,\chi_v,\psi_v)^3\ ,\\ &\qquad v \in S\ ,\ v \neq w\ .\end{aligned}$$

So if we apply the relation (19.6.3)*to σ and $\tau \otimes \chi$, where χ is a quasi-character of $\underline{I}/F^{\times}$ such that $\chi_w = 1$ and the relations (19.6.4) and (19.6.5) are true, we find that the relation (19.16.2) is true for the place w. This concludes the proof of (19.16).

*The relation (19.16.2) still holds if we replace τ by $\tau \otimes \chi$.

§20. An application to quadratic extensions

We first go back to a local situation and introduce a few "ad hoc" notions and remarks.

Let, for one moment, F be a local field and K a separable quadratic extension of F. Let π (resp. σ) be an admissible irreducible representation of $\mathcal{H}_F$ (resp. $\mathcal{H}_K$) the Hecke algebra of the group $GL(2,F)$ (resp. $GL(2,K)$). We assume π and σ to be infinite dimensional. Denote by ω (resp. ω') the quasi-character of $F^\times$ (resp. $K^\times$) such that $\pi(a) = \omega(a)$ for $a \in F^\times$ (resp. $\sigma(b) = \omega'(b)$ for $b \in K^\times$).

Definition 20.1: We shall say that σ is a lifting of π if the following conditions are satisfied:

(20.1.1) $$\omega' = \omega \circ Nr_{K/F} \ ;$$

(20.1.2) for any quasi-character χ of $K^\times$

$$\varepsilon'(s,\pi \times \pi(\chi),\psi_F) = \lambda(K/F,\psi_F)^2 \ \varepsilon'(s,\sigma \otimes \chi,\psi_K) \ .$$

We shall say that σ is a strict lifting of π if, in addition, the following condition is satisfied:

(20.1.3) for any quasi-character χ of $K^\times$,

$$L(s,\pi\times\pi(\chi)) = L(s,\sigma \otimes \chi) \ .$$

As usual ψ_F denotes a nontrivial additive character of F and ψ_K is the additive character $\psi_F \circ Tr_{K/F}$ of K. The factor $\lambda(K/F,\psi_F)$ and the representation $\pi(\chi)$ have been defined in [1]. Actually, $\pi(\chi)$ is nothing else than the representation $\pi(\tau)$ of [1], §12, where τ is the representation of $W_{K/F}$ defined by

$$\tau = \mathrm{Ind}(W_{K/F}, W_{F/F}, \chi) \quad .$$

Within an equivalence it is characterized by the following conditions. For a in $F^{\times}$,

$$\pi(\chi)(a) = \chi(a)\zeta(a)$$

where ζ is the quadratic character of $F^{\times}$ attached to K. If η is a quasi-character of $F^{\times}$, the representation $\pi(\chi) \otimes \eta$ is just the representation $\pi(\chi')$ where $\chi' = \chi \cdot (\eta \circ Nr_{K/F})$. Finally

$$\epsilon'(s,\pi(\chi),\psi_F) = \lambda(K/F,\psi_F)\epsilon'(s,\chi,\psi_K) \quad .$$

If the condition (20.1.2) is satisfied for a choice of ψ_F, it is satisfied for all choices. If there is a lifting of π, it is unique up to equivalence.

There are a number of cases where we can prove the existence of a lifting.

(20.2) Suppose that F is nonarchimedean and that $\pi = \pi(\mu,\nu)$ (with $\mu.\nu^{-1}$ different from α_F and α_F^{-1}), or $\pi = \sigma(\mu,\nu)$ (with $\mu.\nu^{-1} = \alpha_F$). Define

$$\mu' = \mu \circ Nr_{K/F} \;,\; \nu' = \nu \circ Nr_{K/F} \quad .$$

Let σ be the representation $\pi(\mu',\nu')$ if $\mu'.\nu'^{-1} \neq \alpha_K^{\pm 1}$, and the representation $\sigma(\mu',\nu')$ if $\mu'.\nu'^{-1} = \alpha_K$ or α_K^{-1}. Then σ is a lifting of π. For $\omega' = \mu'.\nu'$. Hence the condition (20.1.1) is surely satisfied. On the other hand

$$\begin{aligned}\epsilon'(s,\pi\times\pi(\chi),\psi_F) &= \epsilon'(s,\pi(\chi)\otimes\mu,\psi_F)\epsilon'(s,\pi(\chi)\otimes\nu,\psi_F) \\ &= \epsilon'(s,\pi(\chi\mu'),\psi_F)\; \epsilon'(s,\pi(\chi\nu'),\psi_F) \\ &= (\lambda(K/F,\psi_F))^2\; \epsilon'(s,\chi\mu',\psi_K)\; \epsilon'(s,\chi\nu',\psi_K)\end{aligned}$$

$$= (\lambda(K/F,\psi_F))^2 \ \varepsilon'(s,\sigma \otimes \chi,\psi_K) \quad .$$

This is condition (20.1.2).

(20.3) Keeping the same assumptions, suppose that π is pre-lunitary (unitary in the terminology of [1]). Then we claim that σ is a strict lifting of π.

If $\mu.\nu^{-1}$ is different from α_F, α_F^{-1}, $\alpha_F\zeta$, $\alpha_F^{-1}\zeta$, then $\mu'.\nu'^{-1}$ is different from α_K and α_K^{-1}. Hence $\pi = \pi(\mu,\nu)$, $\sigma = \pi(\mu',\nu')$ and

$$\begin{aligned} L(s,\pi\times\pi(\chi)) &= L(s,\pi(\chi)\otimes\mu)L(s,\pi(\chi)\otimes\nu) \\ &= L(s,\pi(\chi\mu'))L(s,\pi(\chi\nu')) \\ &= L(s,\chi\mu')L(s,\chi\nu') \\ &= L(s,\sigma\otimes\chi) \quad . \end{aligned}$$

Hence our assertion.

If $\mu.\nu^{-1} = \alpha_F$ then $\mu'.\nu'^{-1} = \alpha_K$. Hence $\pi = \sigma(\mu,\nu)$, $\sigma = \sigma(\mu',\nu')$ and

$$\begin{aligned} L(s,\pi\times\pi(\chi)) &= L(s,\pi(\chi))\otimes\mu \\ &= L(s,\chi\mu') \\ &= L(s,\sigma\otimes\mu) \quad . \end{aligned}$$

So all we have to show is that $\mu.\nu^{-1}$ cannot be equal to $\zeta\alpha_F$ if π is preunitary. If it was so we would have

$$\mu = \alpha_F^{\frac{1}{2}}\theta \ , \ \nu = \alpha_F^{\frac{1}{2}}\theta\zeta \ ,$$

where θ is some character of $F^\times$. The representation $\tilde{\pi}$ would be equivalent to the conjugate imaginary of π. This would imply that $\{\bar{\mu},\bar{\nu}\} = \{\mu^{-1},\nu^{-1}\}$ which is easily found to be impossible.

(20.4) Suppose that $F = \mathbb{R}$. Then $K = \mathbb{C}$ and $\pi = \pi(\tau)$ where τ is a two dimensional representation of the Weil group $W = W_{\mathbb{C}/\mathbb{R}}$. The restriction τ' of τ to $\mathbb{C}^\times$ has the form $\tau' = \mu' \oplus \nu'$ where μ' and ν' are quasi-characters. Let σ be $\pi(\mu',\nu')$ or, if $\mu'.\nu'^{-1} = \alpha_{\mathbb{C}}$ or $\alpha_{\mathbb{C}}^{-1}$, $\sigma(\mu',\nu')$. Then σ is a lifting of π.

For if ρ is the representation $\mathrm{Ind}(W,\mathbb{C}^\times,\chi)$ we have $\pi(\chi) = \pi(\rho)$, $\tau \otimes \rho = \mathrm{Ind}(W,\mathbb{C}^\times,\tau' \otimes \chi)$ and

$$\begin{aligned}\epsilon'(s,\pi\times\pi(\chi),\psi_F) &= \epsilon'(s,\pi(\tau) \otimes \pi(\rho),\psi_F)\\ &= \epsilon'(s,\tau \otimes \rho,\psi_F)\\ &= (\lambda(K/F,\psi_F))^2\ \epsilon'(s,\tau' \otimes \chi,\psi_K)\\ &= (\lambda(K/F,\psi_F))^2\ \epsilon'(s,\sigma \otimes \chi,\psi_K)\ .\end{aligned}$$

This is condition (20.1.2). The condition (20.1.1)is obviously satisfied.

(20.5) In addition assume that π is preunitary. Then σ is a strict lifting of π. For, in general, the only case where σ is not a strict lifting is when $\pi = \pi(\mu,\nu)$ and the representation $\sigma(\mu',\nu')$ is defined. It is easily seen that this cannot happen if π is preunitary.

Now let us go back to the global situation. Let F be an $\mathbb{A}$ field and K a separable quadratic extension of F. We denote by $F_{\mathbb{A}}$ and $K_{\mathbb{A}}$ the corresponding rings of adèles and by $F_{\mathbb{A}}^\times$ and $K_{\mathbb{A}}^\times$ the groups of idèles. Let π (resp. σ) an irreducible admissible representation of $\mathcal{H}_F$ (resp. $\mathcal{H}_K$) the Hecke algebra of the group $GL(2,F_{\mathbb{A}})$ (resp. $GL(2,K_{\mathbb{A}})$). We assume that all components of π and σ are infinite dimensional. We shall say that σ is a lifting of π if the following conditions are satisfied. For each place v of F we denote by F_v the corresponding local field and by K_v the F_v algebra $F_v \otimes_F K$. If K_v is a quadratic extension of F_v let w be the unique place of K above v. Then $K_w = K_v$ and we want σ_w to be a lifting of π_v.

If $K_v = F_v \oplus F_v$ let w' and w'' be the two places of K above v. Then $K_{w'} = K_{w''} = F_v$ and want $\sigma_{w'} = \sigma_{w''} = \pi_v$.

If there is a lifting at all, it is unique. Moreover, if ω (resp. ω') is the quasi-character of $F_A^\times$ (resp. $K_A^\times$) such that $\pi(a) = \omega(a)$ (resp. $\sigma(a) = \omega'(a)$) for a in $F_A^\times$ (resp. $K_A^\times$), then

$$\omega' = \omega \circ Nr_{K/F} .$$

<u>Theorem 20.6</u>: <u>Suppose</u> ω <u>is a character of</u> $F_A^\times/F^\times$ <u>and</u> π <u>an irreducible component of</u> $\mathcal{A}_0(\omega)$ <u>(space of cusp forms on</u> $GL(2,F_A)$ <u>transforming under</u> $F_A^\times$ <u>according to</u> ω<u>)</u>. <u>Then there is a lifting</u> σ <u>of</u> π <u>to</u> $\mathcal{H}_K$. <u>Moreover either</u> σ <u>is a component of</u> $\mathcal{A}_0(\omega')$ <u>where</u> $\omega' = \omega \circ Nr_{K/F}$, <u>or there are two quasi-characters</u> μ' <u>and</u> ν' <u>of</u> $K_A^\times/K^\times$ <u>such that</u> $\mu'\nu' = \omega'$ <u>and for all places</u> w <u>of</u> K, <u>the representation</u> σ_w <u>is the infinite dimensional component of</u> $\rho(\mu'_w,\nu'_w)$.

As usual we choose a nontrivial character ψ_F of F_A/F. Then $\psi_K = \psi_F \circ Tr_{K/F}$ is nontrivial character of K_A/K. The letter v always denotes a place of F and the letter w a place of K.

Let S be the set of all nonarchimedean places v of F which do not split in K and where π_v has not the form $\pi(\mu_v,\nu_v)$ or $\sigma(\mu_v,\nu_v)$. Then S is finite. If it is empty we take, for convenience, S to be $\{v\}$ where v is some nonarchimedean place which does not split in K. Let T be the set of places of K which are above a place in S. There is a one to one correspondance between S and T but it is best not to identify the two sets. If w is a place of K not in T we define a representation σ_w of $\mathcal{H}_w$ as follows.

If w is above v which does not split, we take σ_w to be the unique lifting of π_v from $\mathcal{H}_v$ to $\mathcal{H}_w$. It is in fact a strict lifting. Therefore for all quasi-characters $\chi_v = \chi_w$ of $K_v^\times = K_w^\times$ we have the relations

(20.6.1)
$$L(s,\pi_v \times \pi(\chi_w)) = L(s,\sigma_w \otimes \chi_w) ,$$
$$L(s,\tilde{\pi}_v \times \tilde{\pi}(\chi_w)) = L(s,\tilde{\sigma}_w \otimes \chi_w^{-1}) ,$$
$$\epsilon(s,\pi_v \otimes \pi(\chi_w),\psi_v) = \lambda_v^2 \epsilon(s,\sigma_w \otimes \chi_w,\psi_w) ,$$

where ψ_v is the local component of ψ_F, $\psi_w = \psi_v \circ Nr_{K_w/F_v}$ is the local component of ψ_K and $\lambda_v = \lambda(K_w/F_v,\psi_v)$. If w' and w'' are the two places above a place v of F which splits in K, we want $\sigma_{w'} = \sigma_{w''} = \pi_v$. Then if $\chi_v = \chi_{w'} \oplus \chi_{w''}$ is a quasi-character of $K_v^\times = K_{w'}^\times \times K_{w''}^\times = F_v^\times \times F_v^\times$, we have the relations

(20.6.2)
$$L(s,\pi_v \times \pi(\chi_{w'},\chi_{w''})) = L(s,\sigma_{w'} \otimes \chi_{w'})L(s,\sigma_{w''} \otimes \chi_{w''}) ,$$
$$L(s,\tilde{\pi}_v \times \tilde{\pi}(\chi_{w'},\chi_{w''})) = L(s,\tilde{\sigma}_{w'} \otimes \chi_{w'}^{-1})L(s,\tilde{\sigma}_{w''} \otimes \chi_{w''}^{-1}) ,$$
$$\epsilon(s,\pi_v \times \pi(\chi_{w'},\chi_{w''}),\psi_v) = \lambda_v^2 \epsilon(s,\sigma_{w'} \otimes \chi_{w'},\psi_{w'})\epsilon(s,\sigma_{w''} \otimes \chi_{w''},\psi_{w''}) ,$$

where ψ_v is the local component of ψ_F at v, $\psi_{w'} = \psi_v$ (resp. $\psi_{w''} = \psi_v$) is the local component of ψ_K at w' (resp. w''), and $\lambda_v = 1$.

If w is in T above v in S, there are quasi-characters χ_w of $K_w^\times$ satisfying the following conditions. First $\chi_w = \eta_v \circ Nr_{K_w/F_v}$

where η_v is a quasi-character of $F_v^\times$; moreover the order of χ_w and η_v is arbitrarily large; finally, the quasi-character $\omega_v\zeta_v \cdot \chi_w|F_v^\times$ is ramified. (We denote by ζ_v the quadratic character of $F_v^\times$ attached to K_w) . So taking a χ_w of this type, we may assume (cf. [1], Proposition 3.8, p.115) that:

(20.6.3)
$$\begin{aligned}
\pi(\chi_w) &= \pi(\eta_v,\eta_v\zeta_v) , \\
\epsilon(s,\pi_v\times\pi(\chi_w),\psi_w) &= \epsilon(s,\pi_v \otimes \eta_v,\psi_v)\epsilon(s,\pi_v \otimes \eta_v\zeta_v,\psi_v) \\
&= \epsilon(s,\omega_v\eta_v,\psi_v)\epsilon(s,\eta_v,\psi_v) \\
&\quad \times \epsilon(s,\omega_v\eta_v\zeta_v,\psi_v)\epsilon(s,\eta_v\zeta_v,\psi_v) \\
&= \lambda_v^2\epsilon(s,\omega'_w\chi_w,\psi_w)\epsilon(s,\chi_w,\psi_w) , \\
L(s,\pi_v\times\pi(\chi_w)) &= L(s,\tilde{\pi}_v \times \tilde{\pi}(\chi_w)) \\
&= 1 .
\end{aligned}$$

We may also assume that the order m_w of χ_w is larger than the order of ω'_w .

The above conditions are satisfied if the restriction of η_v to the group of units of $F^\times$ is suitable. Therefore there is a character η of $F_A^\times/F^\times$ whose components at each v in S satisfy the above relations. Replacing π by $\pi\otimes\eta$ we see that we may <u>assume</u>:

(20.6.4) <u>If</u> χ_w <u>is an unramified quasi-character of</u> $K_w^\times$ <u>then</u>

$$L(s,\pi_v\times\pi(\chi_w)) = L(s,\tilde{\pi}_v\times\tilde{\pi}(\chi_w)) = 1 ,$$
$$\epsilon(s,\pi_v\times\pi(\chi_w)),\psi_v) = \lambda_v^2 c_w\chi_w(\varpi_w)^{m_w+2n_w}|\varpi_w|^{(m_w+2n_w)(s-\frac{1}{2})} ,$$

<u>where</u> $c_w \neq 0$, n_w <u>is the order of</u> ψ_w <u>and</u> $m_w > 0$; <u>the quasi-character</u> $\omega_v\zeta_v(\chi_w|F_v^\times)$ <u>is ramified</u>.

Now let χ be a quasi-character of $K_A^\times/K^\times$ which is unramified at each place w in T. Then by (19.15)and (19.15.1) $L(s,\pi\times\pi(\chi))$ and $L(s,\tilde{\pi}\times\tilde{\pi}(\chi))$ are entire functions of s, bounded in any vertical strip of finite width if F is a number field. They satisfy the functional equation

$$L(s,\pi\times\pi(\chi)) = \epsilon(s,\pi\times\pi(\chi))L(1-s,\tilde{\pi}\times\tilde{\pi}(\chi)) .$$

But in fact, by the above relations,

$$L(s,\pi\times\pi(\chi)) = \prod_{v\notin S} L(s,\pi_v\times\pi(\chi_v)) = \prod_{w\notin T} L(s,\sigma_w\otimes\chi_w)$$

$$L(s,\tilde{\pi}\times\tilde{\pi}(\chi)) = \prod_{v\notin S} L(s,\tilde{\pi}_v\times\tilde{\pi}(\chi_v)) = \prod_{w\notin T} L(s,\tilde{\sigma}_w\otimes\chi_w^{-1}) .$$

Since the product of all λ_v is one, we see that the functional equation reads

$$\prod_{w\notin T} L(s,\sigma_w\otimes\chi_w) = a(s,\chi)\prod_{w\notin T}\epsilon(s,\sigma_w\otimes\chi_w,\psi_w)\prod_{w\notin T} L(1-s,\tilde{\sigma}_w\otimes\chi_w^{-1}) ,$$

where

$$a(s,\chi) = \prod_{w\in T} c_w\chi_w(\varpi_w)^{m_w+2n_w}|\varpi_w|^{(m_w+2n_w)(s-\frac{1}{2})} .$$

Following step by step, the proof of Theorem 12.2 in [1], we see that either there is a component σ of $\mathcal{A}_0(\omega')$ whose component at $w\notin T$ is σ_w, or there are two quasi-characters μ' and ν' of $K_A^\times/K^\times$ such that $\omega'=\mu'\nu'$ and σ_w is the infinite dimensional component of $\rho(\mu_w',\nu_w')$ for all w not in T. In the latter case define σ_w for w in T to be the infinite dimensional component of $\rho(\mu_w',\nu_w')$ and σ to be the product of the σ_w for all w. Then in both cases we have the functional equation

$$L(s,\sigma\otimes\chi) = \epsilon(s,\sigma\otimes\chi)L(1-s,\tilde{\sigma}\otimes\chi^{-1}) ,$$

for all quasi-characters χ of $\underline{I}/F^{\times}$. On the other hand, we have

$$L(s,\pi\times\pi(\chi)) = \epsilon(s,\pi\times\pi(\chi))L(1-s,\tilde{\pi}\times\tilde{\pi}(\chi))$$

for all quasi-characters χ of $\underline{I}/F^{\times}$.

Also we have

$$\sigma_w(a) = \omega'_w(a)$$

for all w and a in $F_w^{\times}$. Using (20.6.1) and (20.6.2) we arrive at the relation

$$\prod_{v\in S}\lambda_v^2\prod_{w\in T}\epsilon'(s,\sigma_w\otimes\chi_w,\psi_w) = \prod_{v\in S}\epsilon'(s,\pi_v\otimes\pi(\chi_v),\psi_v) .$$

Fix v_0 in S and let w_0 be the corresponding element of T. Choose χ in such a manner that for $w\in T$, $w\neq w_0$, χ_w have the form $\chi_w = \eta_v\circ Nr_{K_w/F_v}$ where the orders of χ_w and η_v are sufficiently large. Then we may assume that the conditions (20.6.3) are satisfied and also that

$$L(s,\sigma_w\otimes\chi_w) = L(s,\sigma_w\otimes\chi_w^{-1}) = 1$$

$$\epsilon(s,\sigma_w\otimes\chi_w,\psi_w) = \epsilon(s,\omega'_w\chi_w,\psi_w)\epsilon(s,\chi_w,\psi_w)$$

for all $w\in T$, $w\neq w_0$.

We then find that

$$\epsilon'(s,\pi_{v_0}\times\pi(\chi_{v_0}),\psi_{v_0}) = \lambda_{v_0}^2\,\epsilon'(s,\sigma_{w_0}\otimes\chi_{w_0},\psi_{w_0}) .$$

Now χ_{w_0} is an arbitrary quasi-character of K_{w_0}. It follows that σ_{w_0} is a lifting of π_{v_0}. This concludes the proof of the theorem.

Bibliography

The reader is assumed to have a knowledge of:

[1] H. Jacquet, R. Langlands, Automorphic forms on GL(2), Lecture notes in mathematics, vol. 114, Springer-Verlag, 1970.

He can usefully complete it by reading:

[2] R. Godement, Notes on Jacquet-Langlands theory, Institute for Advanced Study, Princeton, N.J., 1970.

Although not necessary for the lecture of the present set of notes, the following paper will most certainly throw some light on [1]:

[3] A. Weil, Dirichlet Series and automorphic forms, Lecture notes in mathematics, vol. 189, Springer-Verlag.

We take also this opportunity to indicate to the reader the following article which will be of great help in filling the gaps of the last section of [1]:

[4] M. Duflo and J.P. LaBesse, Sur la formule des traces de Selberg, Ann. Scient. Éc. Norm. Sup. 4° série, t. 4, 1971, pp.193-284.

In §4 and §5 we had occasion to refer to:

[5] Whittaker and Watson, A Course of Modern Analysis, Cambridge, 1962.

Section 19 does not contain any new idea. In particular, it is largely based upon:

[6] R. Godement, Analyse spectrale des fonctions modulaires, Séminaire Bourbaki, 1964/65.

For §6 we have found convenient to refer also to:

[7] A. Weil, Basic Number Theory, Springer-Verlag, 1967.

As well as to:

[8] R. Godement, H. Jacquet, Zeta functions of simple algebras, Lecture notes in mathematics, Springer-Verlag , Vol. 260.

The Artin-Hecke L-functions are discussed in:

[9] R.P. Langlands, On the functional equation of the Artin L-functions, Notes, Yale University (in preparation).

The "philosophy" of L-functions associated with automorphic forms is explained in:

[10] R.P. Langlands, Problems in the theory of automorphic forms, Notes, Yale University.

The following papers discuss the same subject as the present one:

[11] R. Rankin, Contributions to the theory of Ramanujan's function, Proc. Cam. Phil. Soc., 1939.

[12] A. Selberg, Bemerkungen über eine Dirichletsche Reihe, die mit der Theorie der Modulformen naheverbunden ist, Arch. Math. Naturvid. 43(1940) 47-50.

[13] A.P. Ogg, On a convolution of L-series, Inventiones math. 7, 1962.

[14] A. Selberg, On the estimation of Fourier coefficients of modular forms, in Theory of Numbers, Proc. of Symposia in Math., Vol. VIII, A.M.S. 1965.

[15] K. Doi and H. Naganuma, On the functional equation of certain Dirichet series, Inventiones math., 9(1969), 1-14.

.ecture Notes in Mathematics

:omprehensive leaflet on request

ol. 74: A. Fröhlich, Formal Groups. IV, 140 pages. 1968. DM 16,-

ol. 75: G. Lumer, Algèbres de fonctions et espaces de Hardy. VI,) pages. 1968. DM 16,-

ol. 76: R. G. Swan, Algebraic K-Theory. IV, 262 pages. 1968. M 18,-

ol. 77: P.-A. Meyer, Processus de Markov: la frontière de Martin. IV, 3 pages. 1968. DM 16,-

ol. 78: H. Herrlich, Topologische Reflexionen und Coreflexionen. XVI, 6 Seiten. 1968. DM 16,-

ol. 79: A. Grothendieck, Catégories Cofibrées Additives et Complexe otangent Relatif. IV, 167 pages. 1968. DM 16,-

ol. 80: Seminar on Triples and Categorical Homology Theory. Edited B. Eckmann. IV, 398 pages. 1969. DM 20,-

ol. 81: J.-P. Eckmann et M. Guenin, Méthodes Algébriques en Méca- que Statistique. VI, 131 pages. 1969. DM 16,-

ol. 82: J. Wloka, Grundräume und verallgemeinerte Funktionen. VIII, 1 Seiten. 1969. DM 16,-

ol. 83: O. Zariski, An Introduction to the Theory of Algebraic Surfaces. econd Printing. IV, 100 pages. 1972. DM 16,-

ol. 84: H. Lüneburg, Transitive Erweiterungen endlicher Permutations- ruppen. IV, 119 Seiten. 1969. DM 16,-

ol. 85: P. Cartier et D. Foata, Problèmes combinatoires de commu- tion et réarrangements. IV, 88 pages. 1969. DM 16,-

ol. 86: Category Theory, Homology Theory and their Applications I. dited by P. Hilton. VI, 216 pages. 1969. DM 16,-

ol. 87: M. Tierney, Categorical Constructions in Stable Homotopy heory. IV, 65 pages. 1969. DM 16,-

ol. 88: Séminaire de Probabilités III. IV, 229 pages. 1969. M 18,-

ol. 89: Probability and Information Theory. Edited by M. Behara, Krickeberg and J. Wolfowitz. IV, 256 pages. 1969. DM 18,-

ol. 90: N. P. Bhatia and O. Hajek, Local Semi-Dynamical Systems. 157 pages. 1969. DM 16,-

ol. 91: N. N. Janenko, Die Zwischenschrittmethode zur Lösung mehr- mensionaler Probleme der mathematischen Physik. VIII, 194 Seiten. 69. DM 16,80

ol. 92: Category Theory, Homology Theory and their Applications II. dited by P. Hilton. V, 308 pages. 1969. DM 20,-

ol. 93: K. R. Parthasarathy, Multipliers on Locally Compact Groups. , 54 pages. 1969. DM 16,-

ol. 94: M. Machover and J. Hirschfeld, Lectures on Non-Standard nalysis. VI, 79 pages. 1969. DM 16,-

ol. 95: A. S. Troelstra, Principles of Intuitionism. II, 111 pages. 1969. M 16,-

ol. 96: H.-B. Brinkmann und D. Puppe, Abelsche und exakte Kate- orien, Korrespondenzen. V, 141 Seiten. 1969. DM 16,-

ol. 97: S. O. Chase and M. E. Sweedler, Hopf Algebras and Galois eory. II, 133 pages. 1969. DM 16,-

ol. 98: M. Heins, Hardy Classes on Riemann Surfaces. III, 106 pages. 69. DM 16,-

ol. 99: Category Theory, Homology Theory and their Applications III. lited by P. Hilton. IV, 489 pages. 1969. DM 24,-

ol. 100: M. Artin and B. Mazur, Etale Homotopy. II, 196 Seiten. 1969. M 16,-

l. 101: G. P. Szegö et G. Treccani, Semigruppi di Trasformazioni ultivoche. VI, 177 pages. 1969. DM 16,-

l. 102: F. Stummel, Rand- und Eigenwertaufgaben in Sobolewschen umen. VIII, 386 Seiten. 1969. DM 20,-

l. 103: Lectures in Modern Analysis and Applications I. Edited by T. Taam. VII, 162 pages. 1969. DM 16,-

l. 104: G. H. Pimbley, Jr., Eigenfunction Branches of Nonlinear erators and their Bifurcations. II, 128 pages. 1969. DM 16,-

l. 105: R. Larsen, The Multiplier Problem. VII, 284 pages. 1969. 1 18,-

l. 106: Reports of the Midwest Category Seminar III. Edited by S. Mac ne. III, 247 pages. 1969. DM 16,-

. 107: A. Peyerimhoff, Lectures on Summability. III, 111 pages. 69. DM 16,-

. 108: Algebraic K-Theory and its Geometric Applications. Edited R. M. F. Moss and C. B. Thomas. IV, 86 pages. 1969. DM 16,-

l. 109: Conference on the Numerical Solution of Differential Equa- ns. Edited by J. Ll. Morris. VI, 275 pages. 1969. DM 18,-

l. 110: The Many Facets of Graph Theory. Edited by G. Chartrand d S. F. Kapoor. VIII, 290 pages. 1969. DM 18,-

Vol. 111: K. H. Mayer, Relationen zwischen charakteristischen Zahlen. III, 99 Seiten. 1969. DM 16,-

Vol. 112: Colloquium on Methods of Optimization. Edited by N. N. Moiseev. IV, 293 pages. 1970. DM 18,-

Vol. 113: R. Wille, Kongruenzklassengeometrien. III, 99 Seiten. 1970. DM 16,-

Vol. 114: H. Jacquet and R. P. Langlands, Automorphic Forms on GL (2). VII, 548 pages. 1970. DM 24,-

Vol. 115: K. H. Roggenkamp and V. Huber-Dyson, Lattices over Orders I. XIX, 290 pages. 1970. DM 18,-

Vol. 116: Séminaire Pierre Lelong (Analyse) Année 1969. IV, 195 pages. 1970. DM 16,-

Vol. 117: Y. Meyer, Nombres de Pisot, Nombres de Salem et Analyse Harmonique. 63 pages. 1970. DM 16,-

Vol. 118: Proceedings of the 15th Scandinavian Congress, Oslo 1968. Edited by K. E. Aubert and W. Ljunggren. IV, 162 pages. 1970. DM 16,-

Vol. 119: M. Raynaud, Faisceaux amples sur les schémas en groupes et les espaces homogènes. III, 219 pages. 1970. DM 16,-

Vol. 120: D. Siefkes, Büchi's Monadic Second Order Successor Arithmetic. XII, 130 Seiten. 1970. DM 16,-

Vol. 121: H. S. Bear, Lectures on Gleason Parts. III, 47 pages. 1970. DM 16,-

Vol. 122: H. Zieschang, E. Vogt und H.-D. Coldewey, Flächen und ebene diskontinuierliche Gruppen. VIII, 203 Seiten. 1970. DM 16,-

Vol. 123: A. V. Jategaonkar, Left Principal Ideal Rings. VI, 145 pages. 1970. DM 16,-

Vol. 124: Séminare de Probabilités IV. Edited by P. A. Meyer. IV, 282 pages. 1970. DM 20,-

Vol. 125: Symposium on Automatic Demonstration. V, 310 pages. 1970. DM 20,-

Vol. 126: P. Schapira, Théorie des Hyperfonctions. XI, 157 pages. 1970. DM 16,-

Vol. 127: I. Stewart, Lie Algebras. IV, 97 pages. 1970. DM 16,-

Vol. 128: M. Takesaki, Tomita's Theory of Modular Hilbert Algebras and its Applications. II, 123 pages. 1970. DM 16,-

Vol. 129: K. H. Hofmann, The Duality of Compact Semigroups and C*- Bigebras. XII, 142 pages. 1970. DM 16,-

Vol. 130: F. Lorenz, Quadratische Formen über Körpern. II, 77 Seiten. 1970. DM 16,-

Vol. 131: A Borel et al., Seminar on Algebraic Groups and Related Finite Groups. VII, 321 pages. 1970. DM 22,-

Vol. 132: Symposium on Optimization. III, 348 pages. 1970. DM 22,-

Vol. 133: F. Topsøe, Topology and Measure. XIV, 79 pages. 1970. DM 16,-

Vol. 134: L. Smith, Lectures on the Eilenberg-Moore Spectral Sequence. VII, 142 pages. 1970. DM 16,-

Vol. 135: W. Stoll, Value Distribution of Holomorphic Maps into Compact Complex Manifolds. II, 267 pages. 1970. DM 18,-

Vol. 136: M. Karoubi et al., Séminaire Heidelberg-Saarbrücken-Strasbuorg sur la K-Théorie. IV, 264 pages. 1970. DM 18,-

Vol. 137: Reports of the Midwest Category Seminar IV. Edited by S. MacLane. III, 139 pages. 1970. DM 16,-

Vol. 138: D. Foata et M. Schützenberger, Théorie Géométrique des Polynômes Eulériens. V, 94 pages. 1970. DM 16,-

Vol. 139: A. Badrikian, Séminaire sur les Fonctions Aléatoires Linéaires et les Mesures Cylindriques. VII, 221 pages. 1970. DM 18,-

Vol. 140: Lectures in Modern Analysis and Applications II. Edited by C. T. Taam. VI, 119 pages. 1970. DM 16,-

Vol. 141: G. Jameson, Ordered Linear Spaces. XV, 194 pages. 1970. DM 16,-

Vol. 142: K. W. Roggenkamp, Lattices over Orders II. V, 388 pages. 1970. DM 22,-

Vol. 143: K. W. Gruenberg, Cohomological Topics in Group Theory. XIV, 275 pages. 1970. DM 20,-

Vol. 144: Seminar on Differential Equations and Dynamical Systems, II. Edited by J. A. Yorke. VIII, 268 pages. 1970. DM 20,-

Vol. 145: E. J. Dubuc, Kan Extensions in Enriched Category Theory. XVI, 173 pages. 1970. DM 16,-

Please turn over

Vol. 146: A. B. Altman and S. Kleiman, Introduction to Grothendieck Duality Theory. II, 192 pages. 1970. DM 18,–

Vol. 147: D. E. Dobbs, Cech Cohomological Dimensions for Commutative Rings. VI, 176 pages. 1970. DM 16,–

Vol. 148: R. Azencott, Espaces de Poisson des Groupes Localement Compacts. IX, 141 pages. 1970. DM 16,–

Vol. 149: R. G. Swan and E. G. Evans, K-Theory of Finite Groups and Orders. IV, 237 pages. 1970. DM 20,–

Vol. 150: Heyer, Dualität lokalkompakter Gruppen. XIII, 372 Seiten. 1970. DM 20,–

Vol. 151: M. Demazure et A. Grothendieck, Schémas en Groupes I. (SGA 3). XV, 562 pages. 1970. DM 24,–

Vol. 152: M. Demazure et A. Grothendieck, Schémas en Groupes II. (SGA 3). IX, 654 pages. 1970. DM 24,–

Vol. 153: M. Demazure et A. Grothendieck, Schémas en Groupes III. (SGA 3). VIII, 529 pages. 1970. DM 24,–

Vol. 154: A. Lascoux et M. Berger, Variétés Kähleriennes Compactes. VII, 83 pages. 1970. DM 16,–

Vol. 155: Several Complex Variables I, Maryland 1970. Edited by J. Horváth. IV, 214 pages. 1970. DM 18,–

Vol. 156: R. Hartshorne, Ample Subvarieties of Algebraic Varieties. XIV, 256 pages. 1970. DM 20,–

Vol. 157: T. tom Dieck, K. H. Kamps und D. Puppe, Homotopietheorie. VI, 265 Seiten. 1970. DM 20,–

Vol. 158: T. G. Ostrom, Finite Translation Planes. IV. 112 pages. 1970. DM 16,–

Vol. 159: R. Ansorge und R. Hass. Konvergenz von Differenzenverfahren für lineare und nichtlineare Anfangswertaufgaben. VIII, 145 Seiten. 1970. DM 16,–

Vol. 160: L. Sucheston, Constributions to Ergodic Theory and Probability. VII, 277 pages. 1970. DM 20,–

Vol. 161: J. Stasheff, H-Spaces from a Homotopy Point of View. VI, 95 pages. 1970. DM 16,–

Vol. 162: Harish-Chandra and van Dijk, Harmonic Analysis on Reductive p-adic Groups. IV, 125 pages. 1970. DM 16,–

Vol. 163: P. Deligne, Equations Différentielles à Points Singuliers Reguliers. III, 133 pages. 1970. DM 16,–

Vol. 164: J. P. Ferrier, Seminaire sur les Algebres Complétes. II, 69 pages. 1970. DM 16,–

Vol. 165: J. M. Cohen, Stable Homotopy. V, 194 pages. 1970. DM 16,–

Vol. 166: A. J. Silberger, PGL_2 over the p-adics: its Representations, Spherical Functions, and Fourier Analysis. VII, 202 pages. 1970. DM 18,–

Vol. 167: Lavrentiev, Romanov and Vasiliev, Multidimensional Inverse Problems for Differential Equations. V, 59 pages. 1970. DM 16,–

Vol. 168: F. P. Peterson, The Steenrod Algebra and its Applications: A conference to Celebrate N. E. Steenrod's Sixtieth Birthday. VII, 317 pages. 1970. DM 22,–

Vol. 169: M. Raynaud, Anneaux Locaux Henséliens. V, 129 pages. 1970. DM 16,–

Vol. 170: Lectures in Modern Analysis and Applications III. Edited by C. T. Taam. VI, 213 pages. 1970. DM 18,–

Vol. 171: Set-Valued Mappings, Selections and Topological Properties of 2^X. Edited by W. M. Fleischman. X, 110 pages. 1970. DM 16,–

Vol. 172: Y.-T. Siu and G. Trautmann, Gap-Sheaves and Extension of Coherent Analytic Subsheaves. V, 172 pages. 1971. DM 16,–

Vol. 173: J. N. Mordeson and B. Vinograde, Structure of Arbitrary Purely Inseparable Extension Fields. IV, 138 pages. 1970. DM 16,–

Vol. 174: B. Iversen, Linear Determinants with Applications to the Picard Scheme of a Family of Algebraic Curves. VI, 69 pages. 1970. DM 16,–

Vol. 175: M. Brelot, On Topologies and Boundaries in Potential Theory. VI, 176 pages. 1971. DM 18,–

Vol. 176: H. Popp, Fundamentalgruppen algebraischer Mannigfaltigkeiten. IV, 154 Seiten. 1970. DM 16,–

Vol. 177: J. Lambek, Torsion Theories, Additive Semantics and Rings of Quotients. VI, 94 pages. 1971. DM 16,–

Vol. 178: Th. Bröcker und T. tom Dieck, Kobordismentheorie. XVI, 191 Seiten. 1970. DM 18,–

Vol. 179: Seminaire Bourbaki – vol. 1968/69. Exposés 347-363. IV. 295 pages. 1971. DM 22,–

Vol. 180: Séminaire Bourbaki – vol. 1969/70. Exposés 364-381. IV, 310 pages. 1971. DM 22,–

Vol. 181: F. DeMeyer and E. Ingraham, Separable Algebras over Commutative Rings. V, 157 pages. 1971. DM 16.–

Vol. 182: L. D. Baumert. Cyclic Difference Sets. VI, 166 pages. 1971. DM 16,–

Vol. 183: Analytic Theory of Differential Equations. Edited by P. F. Hsieh and A. W. J. Stoddart. VI, 225 pages. 1971. DM 20,–

Vol. 184: Symposium on Several Complex Variables, Park City, Utah, 1970. Edited by R. M. Brooks. V, 234 pages. 1971. DM 20,–

Vol. 185: Several Complex Variables II, Maryland 1970. Edited by J. Horváth. III, 287 pages. 1971. DM 24,–

Vol. 186: Recent Trends in Graph Theory. Edited by M. Capobianco/J. B. Frechen/M. Krolik. VI, 219 pages. 1971. DM 18.–

Vol. 187: H. S. Shapiro, Topics in Approximation Theory. VIII, 275 pages. 1971. DM 22,–

Vol. 188: Symposium on Semantics of Algorithmic Languages. Edited by E. Engeler. VI, 372 pages. 1971. DM 26,–

Vol. 189: A. Weil, Dirichlet Series and Automorphic Forms. V, 164 pages. 1971. DM 16,–

Vol. 190: Martingales. A Report on a Meeting at Oberwolfach, May 17-23, 1970. Edited by H. Dinges. V, 75 pages. 1971. DM 16,–

Vol. 191: Séminaire de Probabilités V. Edited by P. A. Meyer. IV, 372 pages. 1971. DM 26,–

Vol. 192: Proceedings of Liverpool Singularities – Symposium I. Edited by C. T. C. Wall. V, 319 pages. 1971. DM 24,–

Vol. 193: Symposium on the Theory of Numerical Analysis. Edited by J. Ll. Morris. VI, 152 pages. 1971. DM 16,–

Vol. 194: M. Berger, P. Gauduchon et E. Mazet. Le Spectre d'une Variété Riemannienne. VII, 251 pages. 1971. DM 22,–

Vol. 195: Reports of the Midwest Category Seminar V. Edited by J.W. Gray and S. Mac Lane. III, 255 pages. 1971. DM 22,–

Vol. 196: H-spaces – Neuchâtel (Suisse)- Août 1970. Edited by F. Sigrist, V, 156 pages. 1971. DM 16,–

Vol. 197: Manifolds – Amsterdam 1970. Edited by N. H. Kuiper. V, 23[illegible] pages. 1971. DM 20,–

Vol. 198: M. Hervé, Analytic and Plurisubharmonic Functions in Finite and Infinite Dimensional Spaces. VI, 90 pages. 1971. DM 16.–

Vol. 199: Ch. J. Mozzochi, On the Pointwise Convergence of Fourier Series. VII, 87 pages. 1971. DM 16,–

Vol. 200: U. Neri, Singular Integrals. VII, 272 pages. 1971. DM 22,–

Vol. 201: J. H. van Lint, Coding Theory. VII, 136 pages. 1971. DM 16,–

Vol. 202: J. Benedetto, Harmonic Analysis on Totally Disconnected Sets. VIII, 261 pages. 1971. DM 22,–

Vol. 203: D. Knutson, Algebraic Spaces. VI, 261 pages. 1971. DM 22,–

Vol. 204: A. Zygmund, Intégrales Singulières. IV, 53 pages. 1971. DM 16,–

Vol. 205: Séminaire Pierre Lelong (Analyse) Année 1970. VI, 243 pages. 1971. DM 20,–

Vol. 206: Symposium on Differential Equations and Dynamical Systems. Edited by D. Chillingworth. XI, 173 pages. 1971. DM 16,–

Vol. 207: L. Bernstein, The Jacobi-Perron Algorithm – Its Theory and Application. IV, 161 pages. 1971. DM 16,–

Vol. 208: A. Grothendieck and J. P. Murre, The Tame Fundamental Group of a Formal Neighbourhood of a Divisor with Normal Crossings on a Scheme. VIII, 133 pages. 1971. DM 16,–

Vol. 209: Proceedings of Liverpool Singularities Symposium II. Edited by C. T. C. Wall. V, 280 pages. 1971. DM 22,–

Vol. 210: M. Eichler, Projective Varieties and Modular Forms. III, 11[illegible] pages. 1971. DM 16,–

Vol. 211: Théorie des Matroïdes. Edité par C. P. Bruter. III, 10[illegible] pages. 1971. DM 16,–

Vol. 212: B. Scarpellini, Proof Theory and Intuitionistic Systems. VII, 291 pages. 1971. DM 24,–

Vol. 213: H. Hogbe-Nlend, Théorie des Bornologies et Applications. V, 168 pages. 1971. DM 18,–

Vol. 214: M. Smorodinsky, Ergodic Theory, Entropy. V, 64 pages. 1971. DM 16,–

Vol. 215: P. Antonelli, D. Burghelea and P. J. Kahn, The Concordance-Homotopy Groups of Geometric Automorphism Groups. X, 140 pages. 1971. DM 16,-

Vol. 216: H. Maaß, Siegel's Modular Forms and Dirichlet Series. VII, 328 pages. 1971. DM 20,-

Vol. 217: T. J. Jech, Lectures in Set Theory with Particular Emphasis on the Method of Forcing. V, 137 pages. 1971. DM 16,-

Vol. 218: C. P. Schnorr, Zufälligkeit und Wahrscheinlichkeit. IV, 212 Seiten 1971. DM 20,-

Vol. 219: N. L. Alling and N. Greenleaf, Foundations of the Theory of Klein Surfaces. IX, 117 pages. 1971. DM 16,-

Vol. 220: W. A. Coppel, Disconjugacy. V, 148 pages. 1971. DM 16,-

Vol. 221: P. Gabriel und F. Ulmer, Lokal präsentierbare Kategorien. V, 200 Seiten. 1971. DM 18,-

Vol. 222: C. Meghea, Compactification des Espaces Harmoniques. III, 108 pages. 1971. DM 16,-

Vol. 223: U. Felgner, Models of ZF-Set Theory. VI, 173 pages. 1971. DM 16,-

Vol. 224: Revètements Etales et Groupe Fondamental. (SGA 1). Dirigé par A. Grothendieck XXII, 447 pages. 1971. DM 30,-

Vol. 225: Théorie des Intersections et Théorème de Riemann-Roch. (SGA 6). Dirigé par P. Berthelot, A. Grothendieck et L. Illusie. XII, 700 pages. 1971. DM 40,-

Vol. 226: Seminar on Potential Theory, II. Edited by H. Bauer. IV, 170 pages. 1971. DM 18,-

Vol. 227: H. L. Montgomery, Topics in Multiplicative Number Theory. IX, 178 pages. 1971. DM 18,-

Vol. 228: Conference on Applications of Numerical Analysis. Edited by J. Ll. Morris. X, 358 pages. 1971. DM 26,-

Vol. 229: J. Väisälä, Lectures on n-Dimensional Quasiconformal Mappings. XIV, 144 pages. 1971. DM 16,-

Vol. 230: L. Waelbroeck, Topological Vector Spaces and Algebras. VII, 158 pages. 1971. DM 16,-

Vol. 231: H. Reiter, L^1-Algebras and Segal Algebras. XI, 113 pages. 1971. DM 16,-

Vol. 232: T. H. Ganelius, Tauberian Remainder Theorems. VI, 75 pages. 1971. DM 16,-

Vol. 233: C. P. Tsokos and W. J. Padgett. Random Integral Equations with Applications to Stochastic Systems. VII, 174 pages. 1971. DM 18,-

Vol. 234: A. Andreotti and W. Stoll. Analytic and Algebraic Dependence of Meromorphic Functions. III, 390 pages. 1971. DM 26,-

Vol. 235: Global Differentiable Dynamics. Edited by O. Hájek, A. J. Lohwater, and R. McCann. X, 140 pages. 1971. DM 16,-

Vol. 236: M. Barr, P. A. Grillet, and D. H. van Osdol. Exact Categories and Categories of Sheaves. VII, 239 pages. 1971, DM 20,-

Vol. 237: B. Stenström. Rings and Modules of Quotients. VII, 136 pages. 1971. DM 16,-

Vol. 238: Der kanonische Modul eines Cohen-Macaulay-Rings. Herausgegeben von Jürgen Herzog und Ernst Kunz. VI, 103 Seiten. 1971. DM 16,-

Vol. 239: L. Illusie, Complexe Cotangent et Déformations I. XV, 355 pages. 1971. DM 26,-

Vol. 240: A. Kerber, Representations of Permutation Groups I. VII, 192 pages. 1971. DM 18,-

Vol. 241: S. Kaneyuki, Homogeneous Bounded Domains and Siegel Domains. V, 89 pages. 1971. DM 16,-

Vol. 242: R. R. Coifman et G. Weiss, Analyse Harmonique Non-Commutative sur Certains Espaces. V, 160 pages. 1971. DM 16,-

Vol. 243: Japan-United States Seminar on Ordinary Differential and Functional Equations. Edited by M. Urabe. VIII, 332 pages. 1971. DM 26,-

Vol. 244: Séminaire Bourbaki – vol. 1970/71. Exposés 382–399. IV, 356 pages. 1971. DM 26,-

Vol. 245: D. E. Cohen, Groups of Cohomological Dimension One. V, 99 pages. 1972. DM 16,-

Vol. 246: Lectures on Rings and Modules. Tulane University Ring and Operator Theory Year, 1970–1971. Volume I. X, 661 pages. 1972. DM 40,-

Vol. 247: Lectures on Operator Algebras. Tulane University Ring and Operator Theory Year, 1970–1971. Volume II. XI, 786 pages. 1972. DM 40,-

Vol. 248: Lectures on the Applications of Sheaves to Ring Theory. Tulane University Ring and Operator Theory Year, 1970–1971. Volume III. VIII, 315 pages. 1971. DM 26,-

Vol. 249: Symposium on Algebraic Topology. Edited by P. J. Hilton. VII, 111 pages. 1971. DM 16,-

Vol. 250: B. Jónsson, Topics in Universal Algebra. VI, 220 pages. 1972. DM 20,-

Vol. 251: The Theory of Arithmetic Functions. Edited by A. A. Gioia and D. L. Goldsmith VI, 287 pages. 1972. DM 24,-

Vol. 252: D. A. Stone, Stratified Polyhedra. IX, 193 pages. 1972. DM 18,-

Vol. 253: V. Komkov, Optimal Control Theory for the Damping of Vibrations of Simple Elastic Systems. V, 240 pages. 1972. DM 20,-

Vol. 254: C. U. Jensen, Les Foncteurs Dérivés de $\varprojlim$ et leurs Applications en Théorie des Modules. V, 103 pages. 1972. DM 16,-

Vol. 255: Conference in Mathematical Logic – London '70. Edited by W. Hodges. VIII, 351 pages. 1972. DM 26,-

Vol. 256: C. A. Berenstein and M. A. Dostal, Analytically Uniform Spaces and their Applications to Convolution Equations. VII, 130 pages. 1972. DM 16,-

Vol. 257: R. B. Holmes, A Course on Optimization and Best Approximation. VIII, 233 pages. 1972. DM 20,-

Vol. 258: Séminaire de Probabilités VI. Edited by P. A. Meyer. VI, 253 pages. 1972. DM 22,-

Vol. 259: N. Moulis, Structures de Fredholm sur les Variétés Hilbertiennes. V, 123 pages. 1972. DM 16,-

Vol. 260: R. Godement and H. Jacquet, Zeta Functions of Simple Algebras. IX, 188 pages. 1972. DM 18,-

Vol. 261: A. Guichardet, Symmetric Hilbert Spaces and Related Topics. V, 197 pages. 1972. DM 18,-

Vol. 262: H. G. Zimmer, Computational Problems, Methods, and Results in Algebraic Number Theory. V, 103 pages. 1972. DM 16,-

Vol. 263: T. Parthasarathy, Selection Theorems and their Applications. VII, 101 pages. 1972. DM 16,-

Vol. 264: W. Messing, The Crystals Associated to Barsotti-Tate Groups: with Applications to Abelian Schemes. III, 190 pages. 1972. DM 18,-

Vol. 265: N. Saavedra Rivano, Catégories Tannakiennes. II, 418 pages. 1972. DM 26,-

Vol. 266: Conference on Harmonic Analysis. Edited by D. Gulick and R. L. Lipsman. VI, 323 pages. 1972. DM 24,-

Vol. 267: Numerische Lösung nichtlinearer partieller Differential- und Integro-Differentialgleichungen. Herausgegeben von R. Ansorge und W. Törnig, VI, 339 Seiten. 1972. DM 26,-

Vol. 268: C. G. Simader, On Dirichlet's Boundary Value Problem. IV, 238 pages. 1972. DM 20,-

Vol. 269: Théorie des Topos et Cohomologie Etale des Schémas. (SGA 4). Dirigé par M. Artin, A. Grothendieck et J. L. Verdier. XIX, 525 pages. 1972. DM 50,-

Vol. 271: J. P. May, The Geometry of Iterated Loop Spaces. IX, 175 pages. 1972. DM 18,-

Vol. 272: K. R. Parthasarathy and K. Schmidt, Positive Definite Kernels, Continuous Tensor Products, and Central Limit Theorems of Probability Theory. VI, 107 pages. 1972. DM 16,-

Vol. 273: U. Seip, Kompakt erzeugte Vektorräume und Analysis. IX, 119 Seiten. 1972. DM 16,-

Vol. 274: Toposes, Algebraic Geometry and Logic. Edited by F. W. Lawvere. VI, 189 pages. 1972. DM 18,-

Vol. 275: Séminaire Pierre Lelong (Analyse) Année 1970–1971. VI, 181 pages. 1972. DM 18,-

Vol. 276: A. Borel, Représentations de Groupes Localement Compacts. V, 98 pages. 1972. DM 16,-

Vol. 277: Séminaire Banach. Edité par C. Houzel. VII, 229 pages. 1972. DM 20,-

Vol. 278: H. Jacquet, Automorphic Forms on GL(2). Part II. XIII, 142 pages. 1972. DM 16,-